AF890367

THE TRANSFORMATION OF ITALIAN ARMED FORCES IN COMPARATIVE PERSPECTIVE

Military Strategy and Operational Art

Edited by Professor Howard M. Hensel, Air War College, USA

The Ashgate Series on Military Strategy and Operational Art analyzes and assesses the synergistic interrelationship between joint and combined military operations, national military strategy, grand strategy, and national political objectives in peacetime, as well as during periods of armed conflict. In doing so, the series highlights how various patterns of civil–military relations, as well as styles of political and military leadership influence the outcome of armed conflicts. In addition, the series highlights both the advantages and challenges associated with the joint and combined use of military forces involved in humanitarian relief, nation building, and peacekeeping operations, as well as across the spectrum of conflict extending from limited conflicts fought for limited political objectives to total war fought for unlimited objectives. Finally, the series highlights the complexity and challenges associated with insurgency and counter-insurgency operations, as well as conventional operations and operations involving the possible use of weapons of mass destruction.

Also in this series:

The Powell Doctrine and US Foreign Policy
Luke Middup
ISBN 978 1 4724 2565-2

Inside Cambodian Insurgency
A Sociological Perspective on Civil Wars and Conflict
Daniel Bultmann
ISBN 978 1 4724 4305 2

The Failure to Prevent World War I
The Unexpected Armageddon
Hall Gardner
ISBN 978 1 4724 3056 4

Air Power in UN Operations
Wings for Peace
Edited by A. Walter Dorn
ISBN 978 1 4724 3546 0

The Transformation of Italian Armed Forces in Comparative Perspective

Adapt, Improvise, Overcome?

FABRIZIO COTICCHIA
Scuola Superiore Sant'Anna and European University Institute, Italy

FRANCESCO N. MORO
University of Milan-Bicocca, Italy

Routledge
Taylor & Francis Group
LONDON AND NEW YORK

First published 2015 by Ashgate Publishing

Published 2016 by Routledge
2 Park Square, Milton Park, Abingdon, Oxon OX14 4RN
711 Third Avenue, New York, NY 10017, USA

Routledge is an imprint of the Taylor & Francis Group, an informa business

Copyright © Fabrizio Coticchia and Francesco N. Moro 2015

Fabrizio Coticchia and Francesco N. Moro have asserted their right under the Copyright, Designs and Patents Act, 1988, to be identified as the authors of this work.

All rights reserved. No part of this book may be reprinted or reproduced or utilised in any form or by any electronic, mechanical, or other means, now known or hereafter invented, including photocopying and recording, or in any information storage or retrieval system, without permission in writing from the publishers.

Notice:
Product or corporate names may be trademarks or registered trademarks, and are used only for identification and explanation without intent to infringe.

British Library Cataloguing in Publication Data
A catalogue record for this book is available from the British Library

The Library of Congress has cataloged the printed edition as follows:
Coticchia, Fabrizio.
The transformation of Italian armed forces in comparative perspective : adapt, improvise, overcome? / by Fabrizio Coticchia and Francesco N. Moro.
pages cm. -- (Military strategy and operational art)
Includes bibliographical references and index.
ISBN 978-1-4724-2751-9 (hardback) 1. Italy--Military policy. 2. Italy--Armed Forces--Reorganization. 3. Europe--Armed Forces--Reorganization. I. Moro, Francesco N. (Francesco Niccolò) II. Title.
UA740.C69 2015
355'.033545--dc23

2014048309

ISBN 9781472427519 (hbk)

Contents

List of Figures and Tables

Figures

Tables

Acknowledgements

Writing a book on defence in Italy has been as challenging as stimulating. In the past two years, several people contributed decisively by supporting us and providing guidance to go through what is an underexplored issue. It was no surprise that access to the armed forces has not always been easy, given a tradition of limited interaction between research and military practice that dates back to the Cold War. All the more so, we would like to begin by thanking all the military people and institutions that contributed to make this research possible.

In particular, we would like to thank the Brigata Folgore, the Brigata Pinerolo, the Brigata Taurinense, the Stato Maggiore Difesa of the Esercito Italiano, the staff of the Istituto di Scienze Militari Aeronautiche (ISMA) of the Italian Air Force, the NATO Rapid Deployable Corps in Solbiate Olona and the Italian Navy for accepting to cooperate with research. We sincerely hope that this book can be a (little) spark to further cooperation on research, and debate, about military issues in Italy.

The Scuola Superiore Sant'Anna provided financial support to this research through the Research Project SCONFL11FC ('Military transformation in comparative perspective'). Fabrizio Coticchia would also like to acknowledge the Schuman Centre for Advance Studies – Global Governance Programme (Research Area: Europe in the World) of the European University Institute. Fabrizio would deeply thank Matthew Evangelista, Francesco Strazzari, and Ulrich Krotz for the observations and support provided in last two years of research. Francesco Moro would like to acknowledge the Fondazione Bruno Kessler's Centre for International Politics and Conflict Resolution (FBK-CERPIC) in Trento, and its director Filippo Andreatta, for advice and for offering the opportunity to spend three months in the best imaginable place to write the manuscript. At CERPIC, Emanuele Castelli, Tyson Chatagnier, and Moira Osti offered precious comments and support (as well as coffees) in this period. Also, Andrea Lippi and Maurizio Catino, respectively at the University of Florence and at the University of Milan-Bicocca, proved essential in guaranteeing support and release from pressing institutional duties while completing the book.

The authors would also like to thank colleagues and experts who decisively contributed to shape this research. Pietro Batacchi, Luciano Bozzo, Stefano Costalli, Jason Davidson, Giampiero Giacomello, Andrea Locatelli, Vittorio Emanuele Parsi, Stefano Ruzza, commented parts of the book or related papers and gave valuable advice and backing. Marco Di Giulio provided invaluable support, both as a *connoisseur* of organizational theory and of Clint Eastwood's filmography. We also thank Nicola Nasuti for research assistance for Chapter 3.

In France, Etienne de Durand, Corentin Brustlein, Lt. Col. Guillaume Garnier at IFRI provided precious information, especially but not exclusively on sections related to French armed forces. In the UK, Theo Farrell, Malcolm Chalmers and Trevor Taylor helped to better understand the complex transformation of the UK armed forces.

A workshop and a panel held at the International Studies Association Conference (ISA) held in San Francisco in March 2013 gave a spark to this project. We would like to thank in particular Pascal Vennesson, Oliver Daddow, and Eric Dahl, who provided insightful comments and support at a very early stage of the work. At the ISA Conference 2014 in Toronto, Ryan Grauer gave extremely useful suggestion for the development of the research.

An anonymous referee provided insightful comments and suggestions. At Ashgate, we are in debt to Kirstin Howgate and Brenda Sharp for their valuable support.

Fabrizio would like to thank Francesco and Gisella for their never-ending encouragement and love (and 'because you are mine, I walk the line'). Francesco is deeply indebted to Cecilia: Her discrete backing 'tutti i santi giorni' has been a constant source of comfort and joy.

Chapter 1
Introduction

Armed Forces in Europe after the Cold War: Activism and Transformation

'Improvise, adapt, overcome' so the Marine Corps' mantra goes, as popularized by Sergeant Gunny Highway/Clint Eastwood when warning his soldiers about the mutable nature of real war. Preparation to face an extreme, and to some extent unpredictable, set of events is needed, and it is based on the ability to blend learning from the past with a vision of what might be new in the future. If this is difficult for a trained group of individuals, it is even more problematic for complex organizations like the modern military. Not surprisingly, therefore, the transformation of armed forces has been one of the most difficult tasks in human societies, where failure is frequent, problems are abundant, and no easy solutions exist. Accordingly, there is no grand theory of change that has been able to account for the diverse set of factors that drive change, or able to account for its success.

The Western armed forces that emerged from the Cold War had to face radical change in the strategic context, primarily the threats that they had been designed to face (Luttwak, 1996; Freedman, 2006). The process was further complicated by the lack of a clear new mission able to replace territorial defence and thus be the benchmark for strategic planning. First came the idea that the end of the Cold War had brought a 'peace dividend' that could lead to major reductions in personnel and overall budgets in the defence sector (Chan, 1995). As the idea gained ground, the era of humanitarian interventions, that is the use of military force to limit and bring to an end conflicts that caused massive civilian suffering, exhibited a new use of armed forces (Bradol, 2004; Kaldor and Salamon, 2006; Smith, 2006; Bellamy, 2009). The world – or at least the world with which the military must deal – changed again after September 11, when Western armed forces entered a period of full and continuous deployment. The interventions in Iraq and Afghanistan, at least in a second phase, came to epitomize counterinsurgency as a key (possibly *the* key) mission that armed forces should perform (Nagl, 2005; Kilkullen, 2009). Now, with the sunset of counterinsurgency and the re-emergence of regional rivalries – from Eastern Europe to the South and North China seas – some commentators wonder whether the past twenty years and more of interventions and defence policies in the transatlantic alliance have been just a big 'mistake' driven by 'flawed strategic assumptions' (Brown, 2014).

No wonder that the transformation of armed forces followed a tortuous path, and often did not seem perfectly equipped to face the challenge *du jour*. After four decades of relative stability, the missions for the military have constantly changed in the past two decades. On both sides of the Atlantic, the debate on the direction and the depth of transformations has been animated. In the US, the matter is the subject of intense and constant discussion on the extent to which force structure, posture, and doctrine have changed (or should have changed), giving rise to a rich and eclectic body of literature (Cohen, 1996; Arquilla and Ronfeldt, 1997; Biddle, 2004; Gray, 2007; Farrell, Rynning and Terriff, 2013).

In Europe (especially continental Europe), however, the debate has assumed peculiar features. It has often focused on the integration of defence policies, which has for long been one of the objectives of the proponents of a more cohesive European Union. Since the failure of the EDC (European Defence Community) in the 1950s there has been a lively public debate, which intensified after the end of the Cold War, on how that process *should* proceed and why it *was not* going as planned (Bigo, 1996; Missiroli, 2005; Blockmans and Wessel, 2009). The inability to act decisively in the Balkans was clear evidence, the divisions on Iraq (but also, later, on Libya) almost definitive proof. This has continued in recent years, and while calls for European defence follow every crisis, real integration has not yet come about. In 2013 President of the European Council Van Rompuy has acknowledged this trend by pointing out that only a few European countries seize opportunities for military cooperation, despite the advantages 'in terms of savings and sharing of best practices' (EUCO, 2013). The European Council has accordingly called on the Member States 'to deepen defence cooperation by improving the capacity to conduct missions and operations and by making full use of synergies in order to improve the development and availability of the required civilian and military capabilities' (EU Council Conclusions, 2013).

Filling the Gap

Academic research has followed these developments closely, mostly focusing on the political level and on issues such as the creation of a European Security and Defence Policy (ESDP) or a Common Security and Defence Policy (CSDP); and the Europeanization of defence policy (Mérand, 2008) has featured heavily in the debate. From several standpoints, however, the seminal article by Christopher Hill on the 'capabilities-expectations gap' (1993) is still the key interpretative text on both practice and theory concerning ESDP/CSDP. Other works have focused on the political and grand strategic level, adapting mainstream IR theory to some puzzles in European defence (Posen, 2006; Davidson, 2011).

With a few notable exceptions concerning general (Vennesson et al., 2009; Farrell, Rynning and Terriff, 2013; Missiroli, 2013) or more specific operational issues (Farrell, 2010; Catignani, 2012), the changes undergone by single countries have been relatively unexplored. This is a quite interesting feature, given that the armed forces constitute the main agents of the process. An effective shift in European defence planning would in fact require more detailed assessment of the military transformation than has occurred in recent years, with careful consideration of national defence restructuring processes. Hence the transformation of European armed forces should be analysed also from the standpoint of how specific mission requirements and operational experiences have shifted adaptation from politico/strategic-level discourses to organizational change, especially in light of the profound changes that have occurred in the past two decades. This is all the more necessary because, in particular since 9/11, the armed forces of European countries (such as Italy, France and the UK) have been involved in a number of large-scale operations, which created strong pushes for transformation. The premise of the study is that, with access to other major resources (primarily financial) precluded, it is increasingly useful to focus on issues such as adaptation and organizational learning as crucial elements of transformation.

This book is about the change in Italian armed forces since 2001, a year which marked a watershed by ushering in an era of constant military interventions, starting with Afghanistan. The time-span considered into the analysis ends with Libya and its immediate aftermath.[1] Considering a larger period, spanning from the end of the Cold War, might have provided a more complete analysis. However, focusing on the after-2001 makes sense for especially three reasons. First, military interventions of the second period are qualitatively different from those of the Nineties. It is not just the number of troops, but the qualitative features of the missions, mostly due to the severe conflicts (with the possible exception of Lebanon) that were ongoing while Italian operations occurred. These operations were also, at least in the case of Afghanistan, considerably longer than any previously undertaken. This is also linked to interactions with allies, which were both deeper and more continuous. Second, because the overall change in this phase is so large, and as a large driver for this change is to be found in the missions themselves, focusing on a more restricted period of time allows for more economies of scale in the narrative.

This choice also allows maximizing the theoretical value of this research. Theoretically, in fact, the aim of this work is neither to provide a grand theory of what drives military innovation nor to assess all theories that might affect transformation of armed forces – and more broadly defence policy. Others have been already dealing with Italian foreign policy-making and the nexus between domestic and international politics (Brighi, 2013), the interaction among strategic narratives, public opinion and military operations abroad (Coticchia and De Simone, 2014), the Parliament role in defence policy and civil-military relations (D'Amore, 2001). As we show further in Chapter 2, discussion on European defence and security policy already plays an important role in academic literature on defence in Europe. Here, we aim to contribute to the analysis of how complex processes such as defence transformation unfold (Farrell, Rynning and Terriff, 2013). The book attempts to identify the key-elements with which to describe where Italian armed forces (and, to some extent, the French and British ones) stand after more than a decade of intense activity. To this end, the manuscript focuses on new empirical evidence of how the Italian armed forces, compared and contrasted with the French and British ones, have devised their doctrines, their force structure,

1 As for doctrine we include the analysis of the most recent documents (2012–2013) in order to account for the feedbacks of latest operational experiences.

and their budgets. In particular, it wonders how and to what extent factors such as doctrine and strategic framework, budget and resource allocation, force structure and deployment interact in shaping transformation. The book addresses the challenges that these forces have encountered when deployed; how doctrine and assets available were coherent with mission requirements, how they have adapted in the course of missions; the lessons that they have taken (or sent) home. A complex view on how learning happens becomes all the more important for organizations if they are to adapt to new challenges. Recognition of the complexity of the phenomenon should not discard an attempt to provide theoretically-relevant insights. Here, we move into three major directions. First, we show the importance of looking at bottom-up sources of military learning at least in order to better grasp the mechanisms of adaptation (Grissom, 2006). Second, we hint at the importance of mechanisms of inter-organizational learning, as – although the push for change comes from operational experience – the specific paths decided depend not only on solutions provided internally but also on the adoption of practices that come from the experience of allies (the US and NATO to begin with). In other words, to assess transformation from more than a decade of continuous deployment also looking at the continuous interaction with partners allows making a contribution to the theories of military innovation by analysing a somewhat overlooked element. Third, we want to show how important organizational feedback is. By this we mean in particular to look at institutionalization of lessons learned, that is how military organizations recast their institutional design to address the challenges coming from operational experience.

Italy in Comparative Perspective

One of the key features of change in Italian, British and French armed forces is the interaction between two competing pressures. On one hand, there is the above-mentioned activism, with military forces constantly deployed in different theatres. After 2001, Italian armed forces intervened in Afghanistan, Iraq, Lebanon and Libya. The French ones avoided Iraq, and the British ones Lebanon but both engaged in missions elsewhere. On the other hand, this did not happen in a situation of increasing resources devoted to defence. It is true that these missions received funding beyond the conventional defence budget, but the latter had already been considerably decreased in the previous decades (in relative terms) in the three countries. Budget constraints in previous years, more so than after 2001, had determined the structure of forces under this constant pressure to act.

Limited resources have had several consequences. Growing financial constraints gradually affected the sustainability of the military commitment in operations. The capacity to deploy a large number of troops around the world naturally requires an adequate amount of resources, especially for training and equipment. After 2011, the overall number of Italian military personnel deployed abroad was considerably reduced: from 9,000 in 2001 to less than 5,000 in 2013.[2] Tight budget constraints was a crucial determinant of this decline. Moreover, public opinion was increasingly in favour of a reduction in global military engagement amid a context of financial crisis (IAI-CIRCaP, 2014).

Second, the downsizing of financial resources has deeply influenced Italy's unbalanced defence budget. In fact, a decrease in overall resources has worsened the asymmetric relationships among different budget items (personnel, operations and investments). Expenditure on personnel, which accounts for the bulk of the budget (around 65 per cent in recent years), has increased in correspondence with a general reduction of other entries because of its different degree of elasticity.

In France and UK, the pattern has not been particularly different. Well after the financial crisis had affected state budgets and led to thorough revision of defence policy, important military missions were undertaken by choice. Thus, at the same time when important weapons acquisition programmes were being either cut or delayed, for instance in the sector of fighter jets, a mission involving the extensive use of air power was launched in Libya. France's mission in Mali demonstrated its desire to play a role in Africa also through the use of force, notwithstanding the concurrent adoption of austerity measures at home. These factors combine to make military transformation in Europe interesting from both a theoretical and empirical standpoint. Why, more specifically, Italy? The rationale for this choice is two-fold. First, Italy has been long neglected, and little information and analysis are available. This contrasts with the activism of Italian foreign and security policy in terms of participation in military operations abroad (Ignazi, Giacomello and Coticchia, 2012): Italy was the only major European country to take part in all the major operations mentioned above. In the first half of 2014, Italy had 5,000 troops deployed in thirty-three missions around the globe. However, since 2000, its overall troop strength has varied from 6,500 to almost 10,000, with an average of 8,000 units employed abroad. Since 1989, Italy's armed forces have participated in more operations than in the previous 40 years (Coticchia, 2014b). Moreover, Italy is one of the principal contributors to the UN peacekeeping and peacebuilding operations: in 2012 Italy was the twentieth country in terms of military and police forces deployed in UN PKOs, and the number one European contributor.[3] Overall, post-bipolar military dynamism has transformed the static approach adopted by Italy during the Cold War. Often times, the President of the Republic Giorgio Napolitano argued that the armed forces are today the main instrument of Italian foreign policy. As stated by Brighi:

2 Data on Italian military presence abroad are provided by the Ministry of Defence: http://www.difesa.it/OperazioniMilitari/Documents/SIT%20ANNO%202014%20al%2025%20febbraio%202014.pdf.

3 See UN DPKO's website: http://www.un.org/en/peacekeeping/contributors/2012/july12_3.pdf.

> Since the 1990s the country has raised its profile in European affairs, in transatlantic relations and in various arenas of the globe, with its troops, interests and its resources […] Though never an easy or straightforward partner, Italy is a necessary and inescapable one – a country that simply cannot be taken for granted in contemporary international affairs'. (2013, p. 6)

Second, as mentioned above, Italy had to undertake large defence budget cuts earlier and more thoroughly than other European countries. As the impact of the recent economic crisis on defence is still unfolding, focusing on Italy can provide insights into mechanisms that drive resource allocation and the outcomes of specific choices. The need to renovate the armed forces has gradually come to be considered a shared urgent need. Although other European countries are facing the same problems, the Italian crisis is particularly severe. In addition, 'bipolar inefficiencies' still affect the composition of Italy's armed forces, and the sustainability of the entire military instrument is widely perceived to be at risk.

Both financial constraints and the challenges raised by regional crises (from Libya to Sahel, from the Horn of Africa to Eastern Europe) have induced Italian governments to re-think national defence, after years of very limited strategic reflection. In 2014 the cabinet finally announced that a new *Libro Bianco* (White Paper) would be drafted before the end of the year (Coticchia and Moro, 2014). The debate on Italian defence is more needed than ever, as the issues concerning transformation of the national military instrument are politically very significant.

Empirically, the analysis is based on the extensive use of primary sources, starting with interviews with experts, practitioners and military officers mostly conducted in the period from Winter 2013 to Summer 2014. We also relied on archival documents, oral and written accounts of relevant military figures, doctrine manuals, and professional military periodicals. In particular, information on Italy derives from interviews and informal discussions with several senior military officers (including Generals) from the main Brigades involved in military operations abroad (Folgore Paratroopers, Pinerolo mechanized infantry and Taurinense Alpine), staff officers of the General Staff of the Army, officers from Air Force and Navy, staff officers at NATO. Italian official documents from the past 20 years (with a focus on those produced after 2001) have been examined, together with publications of the single services with reference to specific aspects of missions or changes occurring in doctrine, assets, structure. Interviewees also provided grey literature that was useful in reconstructing specific aspects of transformation. A complete list of interviews as well of other material used in the book can be found in the Appendix.

Finally, a note on why we taking into account France and the UK too. This is not a classical comparative analysis designed for standard theory testing. We believe, however, that the introduction of two extra cases is useful for at least two reasons. First, looking at the two major military powers in Europe allows for a clearer picture of where Italy stands. This is specifically the objective of Chapters 3 and 4. Other European powers such as Germany or Spain do not have similar patterns in military activism, for instance, and would not have had the same illustrative function. We believe that such a comparison also allows for some theoretically relevant insight. Though they are very different in terms of historical trajectories and overall resources, in fact, the three countries show similar pressures. This is especially true with reference to the constraints in terms of available funding and to the requirements coming from constant deployment. A comparison will show how, though differences in several fields remain (from the attitude towards autonomous development of doctrine to communication and strategic narratives), some elements indicate how isomorphic pressures also led to a certain convergence, if not in the outcomes at least in the mechanisms that underpin innovation processes. Further research might certainly better grasp causal mechanisms in comparative perspective, perhaps taking into account a wider number of countries within NATO.

Plan of the Book

The next chapter reviews the existing literature and provides the theoretical framework for the book. It sets out the methodological premise of the study, arguing that more in-depth analysis of interaction between macro and meso-level variables is needed, and that a qualitative analysis that includes interviews and analysis of primary sources is required. The chapter lays the book's theoretical foundations, stressing the importance of the configurational analysis typical of institutional studies. The focus is two-fold. First, we identify the main dimensions of transformation and how to assess their co-evolution and coherence. Second, the chapter highlights specific mechanisms and dynamics that might underpin operational and institutional learning.

The third chapter presents an overview of defence transformation in Italy, comparing and contrasting with France and the UK. Through the analysis of major official documents, the first section provides insights into how armed forces and political decision-makers adapted their view of the external environment after the end of the Cold War and how this shaped defence reform in the 1990s and then in particular after 9/11. After illustrating the main doctrinal change (i.e., the way in which armed forces endogenize defence policy change), the chapter provides an analysis of budget evolution in Italy, showing the main differences and similarities with the budgets of the other two countries. The impact of NATO and EU on defence transformation is examined in the last section, which then summarizes the arguments and shows convergences and divergences in the force transformation of the three countries.

Chapter 4 analyses the main military missions undertaken by Italy since 2001. The focus is on the types of forces deployed and the approach adopted. Rather than providing a complete account of the missions, the chapter empirically observes the coherence of force deployment with the type of mission, the process of adapting to the environment based on existing doctrines and on learning on the field, and the channels of communication across strategic levels. For each intervention, the chapter focuses on Italy and provides comparative references when appropriate. Each section contains references to French and British experiences in the same scenario (where possible). This chapter is also based on in-depth and semi-structured interviews with military officers who have participated in planning, missions, and training.

Chapter 5 blends theoretical and empirical insights from Chapters 3 and 4 to provide an analytical summary of the findings, discussing a series of lessons learned from military operations since 2001 in terms of how military organizations deal with change in their environment in regard to challenges and constraints due to shrinking budgets. The first part of the chapter discusses the coherence between doctrine and field experience. The second part observes how institutional change happened in Italy, both looking at the innovations introduced and at the drivers of innovation in the field of lessons learned. Then, a summary of key changes in UK and France is provided. The aim is to assess the institutional impact of the interaction among field experience, doctrine and learning.

Finally, the concluding chapter illustrates the book's main findings and arguments. We summarize the specific insights gained from the interaction among the key elements of transformation, highlighting the main features of the defence model emerged in the past decades, as well as its risks and problems, especially in terms of the sustainability over time.

Chapter 2
Military Transformation: The Quest for Coherence and Adjustment

The explanation of how and why military organizations change has long been a core theme of inquiry in international relations and strategic studies, and over time studies in these disciplines have been enriched by contributions from organizational theory and policy analysis. This chapter has two purposes. The first is to reconstruct the drivers of change in military organizations, and to assess the literature on European armed forces. Starting from what we perceive to be the weaknesses of existing analyses, we show what aspects warrant closer attention and provide a framework within which it is possible to plot an analytical narrative of how the process of change has evolved. Our key argument – which closely mirrors the 'frustration' with grand theories of military innovation expressed in the most recent major book on the subject by Theo Farrell, Terry Terriff and Sten Rynning (2013) – is that understanding change requires looking into the 'black box' of what change actually is, and that an accurate analysis of the process of change is a first essential step that (at least in the case of Italy) has not yet been made.

The chapter proceeds as follows. First, we review different approaches to the interpretation and explanation of transformation in the literature on military innovation. Second, we conduct a critical analysis of existing studies on the transformation of European armed forces. These studies have predominantly taken change as a given outcome and stressed the importance of external factors in explaining it, referring to the evolution of the global security environment or to the process of Europeanization. Here, we stress that such a view is partial because it does not consider how the process of change has actually taken place and overlooks the endogenous dimension of transformation. We then provide a framework in which to single out the most important aspects of transformation and address two specific issues. First, we focus on coherence among the different aspects of defence policy, such as doctrine, budget allocation and force structure, and deployment. Second, we provide some analytical keys with which to understand the mechanisms underlying military adaptation, both within military operations and in regard to change in organizational set-up and doctrine resulting from lessons learned. Finally, we indicate how we proceed in the empirical analysis according to the framework described.

The Sources of Military Innovation: A Review

Studies on military transformation are closely linked to research questions and conceptual tools in the literature on military innovation. Military innovation is defined as 'change that forces one of the primary combat arms of a service to change its concepts of operation and its relation to other combat arms, and to abandon or downgrade traditional missions' (Rosen, 1988, p. 134). The concept of 'force transformation' has emerged more recently as an attempt to describe the changes that armed forces should undergo in response to technological change. The Department of Defense's 2001 *Quadrennial Defense Review* possibly constituted the apex of the process that made the adoption of the idea of a Revolution of Military Affairs (RMA) a centrepiece of American defence policy (Farrell, 2008). 2001 marked both the culmination of a process and the beginning of its demise because the unconventional wars of the 2000s did not follow the conventional pattern envisaged, and they required the redevelopment of capabilities that relied less heavily on technology (Farrell, Rynning and Terriff, 2013). But transformation, in a broader and literal sense, came about nonetheless, and it still makes sense to use the term – especially because we refer to cases in which it has never had such a specific connotation.

Two distinctions are required on what drives, or hinders, innovation in order to make sense of a quite large and heterogeneous body of literature. First, it is possible to distinguish between theories that centre on international factors and theories that focus mostly on domestic/internal ones. Second, some approaches show the impact of more material (and *sensu lato* rational) factors, while others focus on how non-material/ideational (and not strictly rational) elements induce change. These distinctions should be seen as a reading guide, because overlaps of course exist and some missing elements remain. Table 1.1 below summarizes how four sets of theories on military innovation arise from the intersection among these dimensions.

Table 1.1 Sources of military innovation

	Domestic	International
Interest/Rational	*Inter-organizational*	*Competition*
Ideas/Non-rational	*Intra-organizational*	*Emulation*

A first group of theories, mostly linked to Kenneth Waltz's neorealism (1979), argue that military innovation has the functional purpose of adapting to what states perceive to be their relative position in international systems and the consequent threats. National strategies embodied in military doctrine then respond quite coherently to the need to 'secure the state' (Posen, 1984; Resende-Santos, 2007). Posen focuses specifically on the role of civilian leadership (which constantly assesses threats to state power) in promoting military innovation. Change in this case can be induced by civilians teaming up with non-orthodox thinkers within military organizations (the so-called 'mavericks') to overcome organizational resistances and to direct the development of doctrine towards that most fit for enhancing national security (Posen, 1984).

These theories, which may be labelled 'Competition', argue that change is caused by international factors and is motivated by the self-interest of states. They may also be extended in two directions. First, they can incorporate a wider definition of security. In particular, a look at the history of military innovation during the Cold War shows that doctrines and military innovations also served the purpose of strengthening the NATO alliance, often in order to reassure allies (Evangelista, 1986). Second, similar theories explain the origins of innovation at lower strategic levels. If as historian John Lynn put it, 'war is a matter of Darwinian dominance or survival for states' (quoted in Goldman, 2002, p. 42), then it makes sense to think that adaptation to the realities of the field will be swift. Bottom-up innovation has been much less studied (Grissom, 2006), but according to competition-related theories, one may expect that at least armies endowed with a certain amount of resources will be able to cope with new challenges. This does not mean that victory is achieved; rather that adaptation follows the threats arising in the field.

'Emulation' of military innovations in other countries is still linked to the international sphere but it is due to different causal mechanisms. Sociology's new institutionalism underpins this view, which is broadly based on the idea that a state's armed forces adopt the institutions and behaviours of other states' armed forces because they believe that it is 'appropriate' (Di Maggio and Powell, 1983). This means that the quest for a socially-defined legitimacy is as important, if not more so, than actual interests in determining important organizational choices. Militaries thus adopt the models (isomorphism) that they perceive as most prestigious, not the ones best suited to their specific needs (Goldman and Andres, 1999). Two forces often drive this process. First, because environmental ambiguity does not provide clear incentives to how to organize or devise strategies, the adoption of 'off-the-shelf' models held in high esteem is a viable solution (Strang and Macy, 2001; Goldman, 2002). Second, isomorphism is facilitated by the existence of professional networks through which knowledge and practices spread (Goldman 2002; Stone 2004). Starting from a study on the military culture of Irish officers after WWI, Farrell shows how transnational norms on the conduct of conventional warfare and civilian control factor into domestic contexts (Farrell, 2001).

Domestic/internal factors have also been indicated as important in shaping military change. The first type of driver can be broadly related to civilian-military or inter-services interactions. To over-simplify, one may refer to the classic set of explanations centred on bureaucratic politics. Here interests and calculations are the key concern, and different configurations of actors and interests can either hinder or promote innovation. Classic works such as Halperin (1974), for instance, conceive military innovation as profoundly affected by the bureaucratic resistance of military organizations motivated by the need to preserve budget and mission (and identity, see *infra*). Divergent interests among different bureaucracies/services may induce aversion to change because attempts to inhibit change may generate conservative strategies (that preserve sources of power and money) or stop others' innovations that might alter what is perceived as a favourable status quo.

Inter-organizational strife, however, can also lead to innovation (Downs, 1967; Rosen, 1991). In an attempt to gain (and/or sustain) advantage over other services, one branch of the armed forces may foster change in order to gather more resources (Armacost, 1969; Sapolsky, 1972; Cote, 1993). One of the best-known cases is the development of the Polaris submarine-launched missiles by the US Navy, which needed a timely development of a new system to 'compete' for resources with the Air Force, which had hitherto largely dominated the nuclear domain. The Polaris breakthrough also stimulated the Air Force to improve its missiles in response (Sapolsky, 1972). Even explanations about intra-service competition would fit into this category if innovation originates from the desire of one branch to gain more resources than another one.[1] The rivalry between Strategic Air Command and Tactical Air Command in the 1950s and 1960s could fall within this category (Halberstam, 2001; Olsen, 2007).

Innovation can also be the outcome of institutional arrangements that link civilians and militaries in different ways. According to Deborah Avant (1993), innovation in crisis situations is more likely when armed forces face a unified civilian leadership than when they face a divided one. The latter may instead strengthen military autonomy and professionalization because it enables armed forces that want to resist the adoption of certain measures to play two 'principals' against each other. For instance, in the USA, armed forces may use congressional support to oppose a presidential push for change, thereby limiting flexibility in crises. By contrast, armed forces dealing with a unified civilian leadership (as in the UK) should be less autonomous but better able to cope with crises that are not part of their core mission (Avant, 1993).

Organizational features also play a key role. Here organizational culture is crucial in understanding how innovation can come about (Adamsky, 2010). While it is evident that it is not possible to equate organizations as all possessing similar

---

1 Definition of what constitutes the organizational level is notoriously complicated, and here we admittedly force the distinction between inter-organization and intra-organizational. Grissom (2006) distinguishes between inter-service and intra-service models of military innovation by the level at which innovation happens. Here, we distinguish the 'mechanisms' that drive innovation. Thus, intra-service innovation can be driven by inter-organizational strife, once we define organizations not simply as individual services but also as sub-units. In this sense, Grissom's account of Engel's explanation for the rise of cruise missiles as due to an intra-Navy rivalry would also fit into the inter-organizational 'box' (Engel, 1994; Grissom, 2006).

preferences – for instance in terms of resistance to change – variation among different organizational cultures does affect innovation (Kier, 1995; Murray, 1996). The classic case of successful military innovation under Hans Von Seeckt in post-WWI Germany at least to some extent relies on its institutionalized pattern of recognizing mistakes and learning from them (Murray, 1996, p. 314). Attention to the process of identifying lessons, both after and during wars, and the effective reframing of doctrine accordingly, was a key factor in German military effectiveness and an element of distinction from other European countries (Murray, 1992). Organization cultures that rigidly repeat past models – even when they worked – and those that are not responsive to bottom-up feedback because they are rigidly hierarchical can, vice-versa, also hinder innovation (Doughty, 1988; Murray, 1996).

Recognition of culture's centrality should not induce one to think that culture is completely disconnected from other elements. As Stephen Rosen argued, it is a configuration of ideas and positions within the defence bureaucracy, more than ideas on their own, which accounts for innovation, at least in peacetime (Rosen, 1991). 'Ideological struggles' to promote innovation succeed when a specific configuration of agents is in place to allow the promotion of mid-level officers who bring a new theory of victory fit for the next war to higher ranks. The new 'theory' (the ideational element) may focus on weapons systems, organizational change, and doctrine; but its eventual incorporation depends on organizational set-ups able to bring forward coalitions of innovators. The parable of the lambasting of Colonel De Gaulle's views on armoured warfare by superior French officers in the 1930s, contrasted with incentives to German mid-level officers to promote experiences gained in training, is a well-known example of how these factors work in practice (Murray, 1996). Culture, as now classic studies on the subject show, must be clearly embedded in an organization (and its internal practices such as career patterns) if it is to affect concrete behaviours (Schein, 2004). Moreover, culture cannot be considered as a given, and as such impossible to modify. Culture in the military context, too, should be conceived as subject to push and pull factors that modify it, and thus understood and analysed (also) as a process (Hatch, 1993; Farrell, 2001).

Two considerations are in order. On the one hand, the role of technological transformation, in particular related to the impact of ICT and robotics of recent decades, can be largely interpreted through the existing sets of explanations (as in the case of other technological revolutions, beginning with nuclear weapons). However, it might be argued that new technologies – or at least some of them that have relatively low entry costs – create wide windows of opportunity that may lead to 'innovations by default' close to the technological determinism that has often been denied as relevant by scholars of military innovation (Evangelista, 1988). Whatever view is taken on the impact of technology, the procurement process – beyond the more or less justified rhetoric of the 'military-industrial complex' – would be the subject of a different study. Here, we will refer to it in broader terms as part of planning, budgeting and relation to deployment, stressing how the process of incorporation of technological innovations in the armed forces is actually shaped by the interaction of several forces.

On the other hand, some authors have criticized the literature on military innovation as being too closely focused on top-down processes. They instead stress the importance of bottom-up practices in military institutions, which are portrayed as learning organizations (Downie, 1998). The debate on the issue is far from settled, and in an important review article on the topic Adam Grissom has convincingly argued that there is space for investigating cases where bottom-up pushes have successfully led to innovation, through both learning and adaptation (Grissom, 2006; Catignani, 2012). In the next part of the chapter, we argue that by stressing the processual nature of military innovation (and of resistance to innovation), it is possible to observe the weight of both top-down and bottom-up push and pull factors, their interaction, and to provide new insights on how European – and in particular Italian – armed forces have responded to these pressures.

What Has Driven Change in European Armed Forces?

Notwithstanding the rich and eclectic tradition of studies on innovation, most studies on European defence have examined drivers of change by focusing primarily on exogenous factors. Locatelli (2014) usefully distinguished three major sets of externally-driven explanations. A first group of scholars have concentrated on the shifting roles in the transatlantic alliance, for instance arguing that European countries have adapted to the USA's changing preferences on 'division of labour', building capabilities in peace-keeping and peace-building to complement the American superior war-making strengths (Simoni, 2013). The USA, in other words, has been able to induce European armed forces to change in a manner favourable for the advancement of US interests by providing incentives in that direction, as well as by exploiting divisions within the EU member states (thus limiting the extent of ESDP/CSDP), and, simply, by being so far ahead in military power as to make European allies consider alternative solutions to advance their interests in the security realm (Whitman, 2004).

A second set of hypotheses concern what was defined above as 'competition' and consider change to be driven by the 'strategic adjustment' of European countries. With no large army at their borders as in the Cold War, European states have had to deal with a new and heterogeneous set of threats. According to this view, there has been a 'functionalist' adaptation of organizations to their new missions, with national defence budgets providing the major, if not the only, constraint for easily detectable new preferences. Technological change also makes it possible to direct change in the new security environment because it incentivizes the creation of a given type of force. For instance, it allows personnel reduction without a reduction in effectiveness or a potential reduction of the 'footprint' of military missions, doing so through increased munitions precision and the capacity for distant strikes that do not involve a large, if any, troop presence on the ground.

A third set of explanations focus on Europe, arguing that there has been a slow but steady process of 'Europeanization' of decisions also in the realm of foreign and defence policy (Mérand, 2008). The progressive institutionalization of ESDP/CSDP, as well as some steps forward in integration of the defence industry, bears out this thesis, which insists on the gradual disappearance of differences in defence policy due to convergence of norms and interests. A realist version of this view also exists: the ESDP/CSDP would go in the direction of creating a military bloc potentially alternative to the USA in the unipolar moment (Posen, 2006). Policy convergence does not necessarily need to include all European countries; it could be the outcome of strengthened cooperation among smaller groups of them.

The research mentioned above furnished important insights, but also raises some problems that we partially address here. Most of our criticisms are consistent with a large body of strategic studies focused on military innovation (Murray and Millett, 1996) to which we referred above, and with a recent strand in innovation studies that discards macro-level explanations to focus on the actual processes of change (Farrell, Rynning and Terriff, 2013). First, there is a methodological issue. There is no doubt that adjustment has taken place and that today's European armed forces are more suited to facing contemporary challenges than those which emerged from the Cold War. Nevertheless, most studies tend to take the dependent variable 'change' for granted and as self-explanatory, although change is far from linear or unidirectional (a version of the 'dependent variable problem' highlighted by Pierson, 1994). One cannot, in other words, observe a clear pattern that can then be explained with elegant independent variables derived from IR theories. The main problem is that those theories deal mostly with the long run and thus overlook the *process* through which transformations take place. This bias is also due to the fact that several studies on the translation of national visions into operational doctrines have considered contexts where threats were immediately affecting the territorial integrity of nations, such as in the interwar period (Mearsheimer, 1983). But in the absence of vital threats and univocal (or at least clearer) directions to follow, *the process is exactly what matters*; and it is specific mechanisms rather than macro-variables that account better for this phenomenon. As noticed by Evangelista (1988), what is needed is to look at processes, more than just outcomes, in order to incorporate internal and external factors in the analysis.

Moreover, although defence policy is clearly a national policy, so that it makes sense to take countries as basic units of analysis, some studies have shown that considering large aggregates such as national models as the only level of analysis can hinder better understanding of how several factors interact in explaining change (Farrell, 2010). While here we do not deal with unit-level transformation or operations (Ruffa, 2010), we recognize the importance of looking into the 'black box' of national models and accounting for their internal diversity in functioning and change. This also applies to the analysis of cultural factors. Here, change has been explained in light of national traditions or, better, 'strategic cultures' (for two different views, Johnston, 1995 and Gray, 2007). National strategic cultures constitute the cognitive set of constraints and opportunities within which policy can be framed and implemented. While the inclusion of endogenous factors such as culture is an important addition to analyses of armed forces transformation in Italy and Europe, attention should be paid to the specific impact of military culture on different aspects of change, rather than assuming a uniform and coherent influence of national culture on military transformation (for an example in this regard see Farrell, 2008).

A second, and related, point is that existing studies have excluded the role of doctrine at the operational level, considering it to be a by-product of environmental (or politico-strategic) pushes. This would drive – at least among European countries – towards the convergence of defence policies. The problem is two-fold. First, recent research on the theme has cast doubt on effective convergence (Pannier and Schmitt, 2014). Second, even if such convergence has taken place, as we will argue later, the specific causal mechanisms behind it should be further investigated. Legacies and organizational set-ups, for instance, deeply affect the way in which responses to threats are framed. In the past fifteen years, the armed forces of European countries have been faced with challenges not easy to deal with by using the material and ideational resources available. Again, if it is overlooked that transformation is a work-in-progress more than an outcome, it is not possible to factor the operational practice of armed forces, both as an indicator of transformation and as a source of (potential) further change, into the analysis. Armed forces transformation should thus be analysed also from the standpoint of how specific mission requirements and experiences shift ‘adaptation’ from strategic-level discourses to the organizational and operational level, and what feedback then occurs from that level back to institutional change and doctrine (Catignani, 2014).

Finally, several analyses adopt a simplistic view of what a ‘budget’ is. By doing so, they overlook the interaction between budgetary decisions and doctrinal/organizational change. A budget is a constraint to which armed forces (as well as individual services and units within them) are subject, and in order to try and cope with it they pursue different courses of action. Budget cuts can only partially be influenced by armed forces (at least in moments critical for public expenditures), but the way in which armed forces react to cuts is diverse and calls into question their perception of (organizational) survival and success in the long run, as well as critical destinies of individuals and networks in the short run. Moreover, more thorough analysis of budgets is a way to see ‘legacies in action’: legacies do not mean resistance to change per se, but – as the historical institutionalist literature has shown (Pierson, 2004) – they can decisively influence the pattern that change assumes. That is to say, the way in which the budget affects transformation is multi-faceted and cannot be reduced to a dichotomy between budget increase and innovation or vice-versa.

To address these shortcomings, and to devise a novel and more complete way to look at defence transformation in Italy, we suggest considering first the co-evolution of the major dimensions of defence policy, and then the interaction between field experience and military policy.

Assessing the Coherence of Defence Transformation: From Planning to the Field

How does defence policy translate into military practice? How is it possible analytically to describe recent military change and its drivers in major European countries, starting with Italy? A useful way to address these questions and to make sense of change is to examine their main dimensions and take account of three issues alluded to above: doctrine, budget, and deployment. Here, we attempt to redefine the dependent variable of previous studies, to unpack it, to transform it into a process that can be then more thoroughly studied, and finally to provide a first-cut interpretation of its dynamics. In the next paragraph, we will move in the 'opposite' direction and come up with a conceptual framework to analyse how experience in the field affects the organizational structure and institutional learning. As stressed in Chapter 1, the book does not aim to offer a univocal and comprehensive explanation on what has led change, but rather to investigate the process of transformation in incremental steps, and how these interact to create (or to constrain) change.

The first dimension to consider is doctrine. Studies on national cultures of strategy and security have stressed the importance of looking at how defence policy is framed in terms of values and even lexicon (Katzenstein, 1996; Ringsmose and Børgesen, 2011). Military doctrine is a key component of defence policy because it focuses on the main international threats posed to national security and seeks to avoid possible gaps between political goals and military means (Posen, 1984). What we should consider here is what we have called a 'process of translation' from the political to the military level, and its evolution through time – that is, how the military reframes political directions.

The first element identifiable through doctrine is the evolution of the main strategic aims of armed forces. Kier (1995) stresses three possible outcomes for military doctrines in front of transformations. They may be radically altered in a few years; they may remain static despite changes in the strategic scenario; or they may adapt to external conditions maintaining their core values intact. Examination of cultural and discourse-related factors within military organizations is made a very complex task by the secrecy and the low level of information that generally characterize armed forces (Arbuckle, 2006). However, the literature has largely emphasized that official documents are the preeminent items to analyse in order to draw the cognitive maps of actors in the field of defence and security (Johnston, 1995; Berger, 1998).

Doctrine is the institutional operationalization of the cognitive frameworks and preferences of the political and organizational leadership for its executing arms on the battlefield (Pressman and Wildavsky, 1984; Kier, 1995; Russell, 2011). For this reason it is widely regarded as a crucial indicator of institutional learning and innovation. Through doctrine as it emerges from official documents, in fact, it is possible to move beyond the definition of the mission and identify how the means to accomplish it – the armed forces – are envisioned. Doctrinal documents then establish (at least ideally) a force structure and posture coherent with the given requirements through guidance on overall force levels, the distribution of human and material resources, and the assignment of specific tasks.

Budget is the second main dimension. It is the key to understanding the importance of the passage between strategic levels: political decisions on allocation of resources are transformed into specific choices that affect military activity. Here we focus on the evolution of budget allocation at different levels. First, we map general changes in budget allocation among services and among general categories of expenditures, and we show the evolution over time of key indicators on major military acquisition programme. We then analyse the coherence between doctrinal evolution and military budgets. There are two questions here. The first is simply whether the actual allocation of resources reflects the evolution of political and strategic planning. The second is if, as organization theory has shown, trade-offs between the short and long run exist, and how incremental changes, though 'tactically' useful, can inhibit larger 'strategic' transformation (March, 1991). We therefore try to show how different budget allocations reflect different patterns of 'exploitation' and 'exploration' and what constraints exist on choices at a given point in time (March, 1991). In particular, we look at how budgetary decisions strike a balance among the attempts of the organizations involved to maintain their relative positions, the need to address existing challenges, and attention to the long-term horizons and challenges that must inevitably be taken into account when designing military policy.

Deployment is the third aspect. Here, the analysis considers the main field experiences of armed forces in order to focus on two elements. First, how does transformation translate into activity? Decisions on what units and means are to be employed in the field, together with analysis of specific patterns of action of military forces, provide a first-cut view of coherence between doctrine and deployment. Second, reconstructing the deployment of armed forces is essential for observing the diversity of the patterns (Ruffa, 2010). For instance, Italian units deployed in Afghanistan, and to some extent French and British units as well, were faced with new types of operational and tactical challenges. They did not behave homogenously, and this diversity of observed behaviours makes it possible to go beyond aggregate models based on the national level and provide a more fine-grained view of the loci in which transformation took place and preliminary hypotheses on how it happened. How and to what extent do key assets and crucial elements of national doctrine translate to the field of deployment? The next chapters suggest some variance in the relative importance of the different factors in each case. Jointness and multinational interoperability[2], comprehensive approach, digitalisation, net-centric capabilities and force protection, mobility and rapid

2 We conceive 'jointness' as a synonym of the Italian '*interforzizzazione*', which means national inter-services cooperation and integration. With 'multinational interoperability' we define 'combined' cooperation and integration among allies (which obviously can undertake *also joint* operation together).

power projection are the key elements that emerge from the analysis of strategic documents. They help to define the framework of the transformation process that has been altered and modified by financial constraints and field experience in a uninterrupted (and complex) process of change.

Organizing Feedback: From the Field to Doctrine and Force Structure

The second part of the analysis, which will be further discussed in Chapter 5, focuses on how feedback from experience comes about, and on how it affects armed force transformation and military policy more broadly. Of course, because deployment has been constant in the past years, the distinction between the two phases is not necessarily coherent with a chronological narrative and rather it is more useful for disentangling different analytical dimensions. Incidentally, this lack of a clear distinction between peacetime and wartime is one of the elements that distinguish the environment where transformation now takes place from previous analyses of transformation. Somehow, continuous deployment removes one of the typical obstacles to learning in military organizations: that is, the need to acquire knowledge from events that for most members of the organization happen, if they do, 'once in a lifetime' and thus are difficult to interpret (Lampel, Shamsie and Shapira, 2009).

In this regard, it is important to understand how military organizations create, retain and transfer knowledge (Argote and Miron-Spekter, 2011). To better define knowledge creation in the military field we should focus on the identification of lessons that derive both from direct operational experience and from others' experience (which may or may not involve conscious attempts at transfer by others). With 'retention' and 'transfer' we refer to the set of tools and bodies that have the duty to incorporate lessons into a coherent body of validated analysis (lessons learned), to convert lessons into doctrine, and to disseminate the latter to other branches of the organization and to other branches and services (Thomas, Watts Sussman and Henderson 2001; Catignani, 2014).

Two dimensions in particular require treatment with reference to the learning process. The first relates to 'adaptation', by which is meant the change in lower-level practices that constitute military activity (Catignani, 2014). The case of the Afghan intervention is the most interesting because of its length and complexity. Here, a process of continuous change took place both as the challenge evolved and as armed forces adopted different doctrines at the strategic, operational, and tactical levels. Examination of the variety of solutions adopted, or at least the variety in how similar solutions were actually implemented, makes it possible to identify different patterns of adaptation. As recent studies have shown, empirical observation of operational experience can help in disentangling the elements that facilitate learning, with particular regard to organizational cultures, institutional set-ups related to how the various units deployed operate, and intra-organizational relations (Ruffa, 2010; Russell, 2011; Catino and Patriotta, 2013). For instance, analysis of national deployment highlights a massive amount of learning and adaptation regarding the protection of forces in contemporary warfare. Indeed, new equipment and approaches have been developed in the IED's counter-warfare across Afghanistan, Iraq and also Lebanon.

The second dimension is organizational learning – defined as a broader process of transformation that leads to 'the institutionalization of new structures, processes, routines and, most importantly, new conceptual and normative constructs within the organization as a whole' (Catignani, 2014, p.31). The multiple commitments of the past decades make it possible to observe organizational learning at two levels: the first is the most classic one, that is, from the field to the organization, the second is the cross-theatre of operations. At the first level, in order to assess the extent to which learning takes place, and if it remains 'compartmentalized' or flows through the hierarchy of the organization (Jacobides, 2005), we look at the formal structures that exist or that have been created or reshaped in order to retain and transfer knowledge within military organizations. The lesson-learning process in particular is a crucial subject of analysis. Consistently with recent analyses (Ucko, 2008; Catignani, 2014), we focus on how the individual services and the armed forces as a whole have institutionalized the lessons of more than a decade of continuous operational experience.

At the second level, we can look at how lessons learned from operational experience have been able to 'travel' (mostly through previous institutionalization) across missions. Here, an important aspect concerns which lessons and experiences can be effectively applied in arenas different from those from which they are derived. More broadly, learning is about methods as much as, if not even more than, about specific contents. Several scholars have pointed out that organizations can either experience difficulties in interpreting experience or make mistakes in applying what they learn from specific experiences (March, 2010). Analysing lessons learned also means investigating which lessons have been retained, incorporated into doctrine and force structure, and how they may actually hinder effectiveness in the future. This is all the more true in the field of military innovation, where 'winning the next war' does not mean having been the best at fighting the last one. We will distinguish between learning which is context or mission-specific and a broader idea of learning as an institutional asset that shapes how armed forces review their experience and transform it into shared knowledge and a guide for action.

When assessing the extent of institutional learning, the interplay between exogenous and endogenous factors is particularly important for understanding the extent and mechanisms of 'learning from others' (Greve, 2003; 2005). The first (exogenous) element that matters is the 'infectiousness' of an innovation, defined as the product of its availability to others, of their interpretation of the innovation as a success, and of the status of the innovating organization. The second (exogenous) element is social proximity, defined as the existence of social ties that create trust among different actors which are then more willing to adopt solutions that they have not devised. Interpretation of other organizations as similar in terms of structures, culture and interests facilitates the process as well, which is why transfer within existing alliances should be smoother. Third, the (endogenous) susceptibility of organizations that learn is important. This is defined in terms of motivations to learn (because of previous poor performance or inactivity in a given field), capabilities such as size, age, slack and absorption routines. Size has a complex effect on learning. On the one hand, size makes it possible to diminish the risks and relative costs of adopting a new practice. On the other, size can correlate with burdensome procedures and large bureaucratic resistances to change. Multi-unit organizations (like the military) can also exhibit very different patterns of learning, with the organizational core less subject to change and other units more responsive to stimuli from other organizations. A similar dual pattern can be found on analysing the effects of age, a classic indicator of the institutionalization (Huntington, 1968) of an organization. Age can be either an element of resistance or an important capability. The availability of excess resources (organizational slack) also creates both incentives and resistance to innovation (Evangelista, 1988), because availability can either be an opportunity for change or reduce the motivation for it (a version of the so-called winner's curse). Finally, an organization's absorption capabilities, understood as the set of cognitive structures such as prior knowledge (in terms of amounts and intensity) and structures of communication between the organization and the environment and among units of the organization (Cohen and Levinthal, 1990), play a central role in explaining if and how learning takes place.

What Comes Next

The empirical analysis in the following chapters has a double pattern. The next chapter broadly corresponds to research questions on the coherence between two different dimensions of defence policy: doctrine and budget allocation. Chapter 4 provides insights into how doctrine and assets performed in the field by looking at the military operations of Italian (and French and British) troops in Afghanistan, Iraq, Lebanon and Libya. Most of the analysis is devoted to the military mission in Afghanistan because of the length and the magnitude of the transformation of armed forces that occurred in Central Asia. Chapter 5 summarizes the findings of the previous empirical investigation, focusing specifically on the main results regarding the interplay among doctrine, budget and deployment. Devoting attention also to the comparative perspective, the chapter first assesses the process of adaptation that occurred at operational level (from planning to the field) and then illustrates the institutional set-up and how learning happens. Finally the main assets of the Italian force structure are highlighted, with critical discussion of how procurement is related to experience on the ground. In Chapters 4 and 5, we will focus mostly on where, when, and how change has effectively happened, leaving to the concluding chapter the task of summarizing the main continuities and persisting, or new, problems.

Chapter 3
Force Transformation: Italy in Comparative Perspective

According to a former Minister of Defence and Chief of Staff, Giampaolo Di Paola (2012), the Italian armed forces have been able to renovate and adapt in order to face new challenges and responsibilities in an altered international scenario. The academic literature has never assessed how and to what extent such transformation has really occurred.

The aim of this chapter is to furnish an overview of defence transformation in Italy, comparing and contrasting it with that in France and the UK. The United Kingdom and France are the European countries that devote the largest amount of resources in absolute terms to military expenditures. Despite all due national differences, there has been a common activism in the restructuring of forces to better deal with changes in the security environment. All countries have been deeply involved in Afghanistan, which has been the most dangerous and complex intervention carried out by European forces since WWII.

While significant differences among countries (such as recent historical events or the roles played by nuclear forces for France and the UK) hinder a thorough comparative analysis, the shared efforts towards organizational restructuring and the revision of long-held beliefs and institutions allow useful comparison to be made among Rome, London and Paris.

The first section of the chapter focuses on the transformation of the strategic framework and doctrine. Through analysis of major official documents, the section provides insights into how armed forces and political decision-makers adapted their view of the external environment after the end of the Cold War and how this shaped defence reform in the 1990s, and then in particular after 9/11. The section considers major doctrinal change (i.e. the way in which armed forces endogenize defence policy change). Then, we examine budget evolution in Italy, showing the major differences and similarities with respect to the budgets of the other two countries. Besides being important per se, the analysis of budgets yields two further pieces of information. First, it serves to check the coherence between means and ends, that is, between the guidelines set by defence policy and the actual allocation of resources. Second, it shows the constraints to which military organizations must adapt their preferences, thus furnishing insights necessary for the interpretation of organizational choices. Subsequently, we consider the impact of NATO and EU on defence transformation, stressing the relative prominence of the former institution (and, relatedly, of the US) as the key provider of cognitive resources (especially in the case of Italy) and as the key reference for planning and action. The final subsection summarizes the arguments and shows similarities and differences in the three countries' force transformation. The Appendix provides an overview of Italy's main military procurement programme.

Setting the Context: The Evolution of Italian Strategic Framework and Doctrine

The End of the Cold War and the Beginning of the 1990s

The constraints imposed by the Cold War hindered an autonomous Italian defence policy for decades. Indeed, the deployment of troops abroad occurred on very few occasions, and the national armed forces were inadequately structured to undertake complex international military operations. Their main task was simply to protect NATO's south-eastern border from a probable Soviet invasion.

The political, cultural and organizational heritage of the dramatic collapse of the Italian armed forces during WWII had profoundly shaped a prudent, or even hostile, national attitude towards the military instrument (Ignazi, Giacomello and Coticchia, 2012).

During the Cold War, also because of a substantial lack of military actions, public strategic analysis was very limited. Before 1989, Italy released only two major strategic official documents: the *Libro Bianco della Difesa 1977* (Defence White Paper) and the *Libro Bianco della Difesa 1985*.

The first *Libro Bianco*, which explicitly restricted the use of force only to defend national institutions, distinguished between 'war' functions (e.g. halting external aggression, guaranteeing territorial integrity, etc.) and 'peace' functions (e.g. controlling the frontiers and protecting vital infrastructures).

The second *Libro Bianco* prompted debate on a new role for the national military instrument. For the first time, Italian defence policy introduced the concept of 'joint operations'. The White Paper envisaged a new rapidly deployable force and identified the main weaknesses in the military structure (e.g. anti-aircraft or anti-tank missiles, etc.).

The *Libro Bianco della Difesa 1985* was drafted after Italy's participation in the Lebanon peacekeeping operation of 1982. The mission in Beirut was a crucial but isolated event for Italian defence policy, which was still strongly affected by Cold War logic – as evidenced by analyses of the political debate at that time (Calossi, Calugi and Coticchia, 2013). Moreover, Italy's significant role in the same geographical area in 2006 (after the Israel-Hezbollah war) well illustrates the noteworthy evolution of Italian defence policy.

Whereas in 1986, when Libya launched two missiles toward the island of Lampedusa after the US attack against Gaddafi, Rome was averse to employing armed forces, in March 2011 Italy was actively involved in air strikes against Libya as part of the NATO Unified Protector operation. In sum, it was the end of the bipolar confrontation that allowed Italy's effective military dynamism.

Although foreign and defence policy played only a marginal part in recent electoral campaigns, and several studies emphasized how overall indifference of the Italian public towards foreign and military policy existed and persisted after the end of Cold War (Battistelli and Isernia, 1991; Battistelli, 2004), notable changes have occurred in these domains in the past two decades. After the fall of the Berlin Wall, Italy started constantly to deploy armed forces abroad, becoming a security provider following several years of inaction (Carbone, 2011; Giacomello and Verbeek, 2011). Starting from the Nineties, political leaders and public opinion have conceived the military instrument as a key asset in Italian foreign policy (Bellucci, 1998; Battistelli, 2004). Despite the considerable evolution, some elements of continuity emerge. The notion of Italy as a 'pacifist country', which was shared during the Cold War by all the main Italian political cultures and parties (D'Amore, 2001), has not been altered by the end of the bipolar era. Indeed, the bipartisan consensus on national military interventions has been constantly based on the framework of 'peace missions' (Ignazi, Giacomello and Coticchia, 2012; Coticchia, 2013). Moreover, multilateralism is still a fundamental guideline for Italian foreign policy (Ratti, 2011; Battistelli et al., 2012), and decision-makers consider a multilateral framework to be the necessary premise for Italian commitment to a military intervention – as illustrated by the recent war in Libya. Brighi (2013) illustrates the growing attention devoted by the literature towards the crucial link between national identity (and more broadly norms, beliefs and values) and the state's external projection. After the end of the bipolar era Italy has constantly portrayed itself as an 'international peacekeeper' in the global arena (Ignazi, Giacomello and Coticchia, 2012). The complex interaction between 'role', 'rank' and 'status' has traditionally affected Italian foreign (and defence) policy (Santoro, 1991) Some scholars have specifically emphasized the role of international prestige as the explaining variable of the Italian contribution to US-led interventions (Davidson, 2011).

On the whole, the debate on the drivers of Italy's post-Cold War military dynamism is broad and controversial. Notwithstanding wide divergences of opinions, however, there is consensus in the literature on the extent to which the Italian military instrument changed after the Cold War. Its evolution is well illustrated by Italy's involvement in multilateral and multinational operations abroad, with an average of 8,000 units employed abroad in the post-Cold War era. At the beginning of February 2013, Italy was deploying more than 5,500 men and women in 26 missions across 27 countries/areas,[1] with Afghanistan, Kosovo, Lebanon and the Horn of Africa being the major missions (Ministry of Defence, 2012).

The document that shaped the structure of the Italian armed forces in the 1990s in terms of their tasks, aims and organization was the *New Defence Model* (1991), in which power projection capabilities became the key feature of defence

1 See the MoD website: http://www.difesa.it/OperazioniMilitari/Pagine/RiepilogoMissioni.aspx.

policy.[2] The new strategic concept was based on the constant deployment of troops abroad in order to deal with new threats and menaces, 'spreading peace' and promoting stabilization across regional and global crises. A quantitative reduction of personnel (and reforms intended to enhance the role of volunteers), qualitative improvements, flexibility, readiness, and a joint approach were all necessary to support the evolution of the military instrument.[3]

Throughout the decade, there was substantial continuity – at least in terms of narratives – on the defence model to adopt. It is apparent in the document *New Forces for a New Century* (2001), which illustrated the main lessons learned after a decade of national military interventions abroad. This strategic document emphasized the particular features of the 'national way to peace operations' (p. 49): combined use of diplomacy, force and intelligence; a constant focus on reconstruction and development; a proportional use of force for deterrence and protection; the effectiveness of communication; quality of training; and readiness. Specific Italian military tools are now key assets in contemporary operations,[4] such as the *Carabinieri* (a military police force), which have been regularly employed in key roles in operations abroad (Coticchia and Giacomello, 2009).

Several reforms were enacted starting from the 1990s. Law 25/1997 reformed the Chiefs of Staff (*Stato Maggiore della Difesa*) in order to enhance cooperation among forces and remove the separation that had constantly affected the Navy, Army and Air Force in the bipolar era. Moreover, the Chairman of the Joint Chiefs of Staff increased his power, altering his previous status of a *primus inter pares*. However, several crucial aspects, such as personnel management, still remained in the hands of each service. The *New Defence Model* also paved the way for transformation of the Italian forces into all-volunteer services. With Laws 331/2000 and 226/2004 (and the following legislative decrees), Italy suspended conscription and has had an all-volunteer force (open to women) since January 2005.

While these innovations radically altered the structure of the armed forces, primarily by increasing the professionalization of Italian soldiers, economic, cultural and organizational legacies still deeply affect that evolution. Some contradictions in the 'new' Italian defence policy can be emphasized. First, only a limited part of active personnel is effectively deployable abroad: only around 30,000 out of 190,000 units can engage in military operations abroad (Bellinzona, 2007). This is (largely) a legacy from the Cold War, when Italian forces were not even conceived as tools of foreign policy

2 While the *Libro Bianco* 1985 was influenced by the intervention in Lebanon (1982–1984), the *New Defence Model* was significantly affected by Operation Desert Storm in Iraq, where Italy carried out the first air strikes since WWII.

3 *New Defence Model*, 1991.

4 Four types of military operations were mentioned: 'traditional peace operations'; 'police operations in support of peace interventions'; 'stabilization missions', and 'humanitarian interventions'. The document advocated massive economic investments because of the deterioration of equipment after several years of intensive operations.

(Ignazi, Giacomello and Coticchia, 2012) and were often seen as means to maintain employment levels (Panebianco, 1997). Second, and relatedly, a disproportionate share of the Italian defence budget is still devoted to personnel (see below). Third, Italy, while insisting on its role as a provider of security through the deployment of armed forces abroad, has also maintained the tradition of deploying troops for 'domestic operations', such as those against terrorists, organized crime and illegal migration (Coticchia and Giacomello, 2009). This tendency did not disappear: somehow paradoxically, while *Carabinieri* have been employed in the Middle East or in the Balkans, Army units have been 'used' for police tasks for which they were not, obviously, originally trained.

Transformation after 9/11

Considering how much strategic debate has occurred in the past decade, it is somewhat surprising that the most recent Italian major strategic document, the *Libro Bianco 2002* (White Paper), was released only a few months after 9/11.[5] The transformed global scenario was the strategic premise of the document, whose aim was to adapt the Italian defence model to new threats by developing an effective and efficient military instrument. Four main 'missions' were identified: the defence of 'vital interests' (2002, p. 65) against external aggression; protection of the Euro-Atlantic area by contributing to NATO's collective defence; crisis management to maintain peace and international stability; and the protection of national institutions. The multilateral framework (UN, EU, NATO, and OSCE) still represented the basis for Italian defence policy in the new century. Like previous ones, this document called for additional investments in defence policy in order to close the gap with respect to European and Western allies, and it insisted on the need to 'harmonize the national budget' (2002, p. 12) in conformity with a Defence Model based on a 190,000-strong all-volunteer force and the requirements of power projection.

In regard to the renewed structure of the armed forces, the White Paper reviewed all the main reforms that had shaped Italian defence in the post-Cold War era, itemizing the key elements of national military involvement in future operations: rapid and flexible logistic support; readiness; standardization of equipment; sustainability; interoperability; joint coordination; and out-sourcing. Owing to the transformed global security environment, new military programmes should above all guarantee naval and aerial mobility, C4 (Command, Control, Computers and Communication) and C3I (Communication, Command, Control and Intelligence) capabilities, precision firepower, air defence, protection of forces, electronic warfare capabilities, maritime surveillance, and reconnaissance through unmanned vehicles.

5 The Renzi government, in office since February 2014, has recently announced that a new White Paper will be drafted before the end of 2014. See *Partito Democratico* (Democratic Party) – PD (2014).

After the publication of the White Paper, other documents contributed to shaping Italian military doctrine. The aim of the *Chief of the Italian Defence Staff Strategic Concept* (2005) was to provide: 'a strategic vision of the context of military security, to define the conceptual reference points that will guide the process of continued transformation of the military, and, finally, to indicate the general goals to be accomplished in the short, medium and long terms' (2005, p. 5). The document confirmed the same four missions as identified by the *Libro Bianco 2002*, highlighting the main geographical areas of strategic interest.[6] The *Concept* focused on a 'holistic approach' (2005, p. 31) to the undertaking of a wide spectrum of multidimensional operations, which should be linked with political, diplomatic, economic, social, legal and informational initiatives. The document stressed the need to adopt innovative concepts of forces employment: 'As an example, in combat situations the emphasis will no longer be on the quantity but on the quality and suitability of the forces in order to achieve superiority through the joint and combined application of knowledge, speed, precision and efficacy of engagement' (2005, pp. 50–51). The main innovation, compared with the *Libro Bianco 2002*, was the definition of 'combat situations' and their specific military requirements. Interoperability, civil-military cooperation, flexibility and technological advantage over the adversary were conceived as fundamental factors in military missions abroad. The armed forces should be essentially expeditionary, joint and net-centric, combining readiness with high capabilities in surveillance and C2. Moreover, military forces in the new century should be constantly 'trained and culturally prepared to operate in synergy according to effect-based doctrines' (2005, p. 11).

In order to prevent mission creep and to deal with possible asymmetric threats, the document emphasized the need to configure the contingent with advanced combat capabilities: 'The fundamental operational capabilities are: C4-ISTAR,[7] deployment and mobility (rapidity of intervention and action), precision and efficacy of engagement, survival and protection of forces, logistical sustainability and general support' (2005, p. 51). Moreover, three general requirements for enabling capabilities were designated: timely availability of forces (mainly through the rotation of troops); joint integration and combined interoperability; and interdisciplinary action (also among services and ministries). Within this process of transformation, the document recognized the following as medium and short-term goals: Joint Surveillance and Command System (JSCS); a Joint Force Command for net-centric interaction; strengthening (in terms of numbers, full operational capabilities and joint command) of special forces; availability of NATO and European Response Forces; enhanced capabilities in key sectors like CIMIC (civil-military cooperation); detection of NRBC (nuclear, radiological, biological and chemical) weapons; homeland security; MSU

6 They are: 'NATO, the EU, the Balkans, Eastern Europe, the Caucasus, North Africa and the Horn of Africa, the Near and Middle East and the Persian Gulf' (2005, p. 26).

7 Command, Control, Communications, Computers, Information/Intelligence, Surveillance, Targeting Acquisition and Reconnaissance.

(Multinational Specialized Units); amphibious projection and strategic mobility (air and maritime). Finally, specific attention was devoted by the *Concept* to the need for the acquired capabilities to be gradually integrated with existing ones.[8] This should enhance operational readiness, the integration with multinational systems and equipment, and expeditionary and net-centric capabilities.

Investire in Sicurezza (Investing in Security), released in 2005, supplemented the *Concept* by illustrating the requirements for innovative long-term planning. The risks related to an unbalanced budget (with a preponderance of costs devoted to personnel to the detriment of investments in programme and operating costs) were the main challenges for Italian defence because they weakened the 'usability' (2005, p. 21) of the military instrument. The document confirmed the primary role of force multiplier capabilities linked to technological advancement, the deployability of special forces and mobility, and it again envisaged three possible scenarios of future operations: 'defence of the national territory', 'rapid-response multinational operations', and 'decisive missions'[9] (2005, p. 11). Specific attention was paid to the creation of air, land, maritime and joint expeditionary task forces, such as the Joint Surveillance and Command System – JSCS (a joint net-centric system), the highly flexible and technological Land Integrated Brigade, and the approach defined as 'Joint Forward from the Sea' to support a maritime power projection through 'coalition sea-basing' (i.e., deploying at sea all the resources provided for operations abroad).

Net-centric capabilities were the main concern of another major strategic document: *Net-Centric Transformation: The Future of Multinational and Interdisciplinary Interoperability* (2005). The paper identified Network Centric Warfare (NCW)[10] as the key factor in promoting the transformation and innovation of the Italian military because contemporary military operations require a shared situational awareness that can be significantly improved by the net-centric doctrine. For the same reason, the development of new technologies (drones, integrated systems of command and control, digitalization of military manoeuvres, satellite communication) and the adoption of a 'joint' approach were conceived as vital for the development of the military instrument.

The most recent document released by Italian Defence is the *Ministerial Directive* (2013), which was a review of the overall reform proposed by the technocratic government led by Mario Monti since December 2011.[11] According to the document, despite the mounting constraints posed by the economic

8 The document defined this process as: 'operating while transforming' (2005, p. 66).

9 'Decisive missions' included (prolonged) crisis management interventions, operations to ensure NATO collective defence, etc.

10 NCW was defined by the document as the combination of cultural and technical elements that give (information) superiority (2005, p.3).

11 It is worth noting that, except for the *Libro Bianco 2002*, all post-2001 Italian strategic documents have been drawn up with the decisive contribution of Adm. Giampaolo Di Paola (as Chief of the Staff in 2005 and as Minister of Defence in 2013).

crisis, widespread regional and global instability still entailed constant military involvement in multilateral stability operations abroad.

While preserving key capabilities (such as effectiveness, interoperability, technological edge, and protection of personnel), Italian defence should prefer quality over quantity (2013, p. 9), assuring power projection and the sustainability of forces even in a context of economic crisis. Therefore, 'Smart Defence' and 'Pooling&Sharing', which are concepts widely shared in the NATO and EU framework, fit perfectly with the existing national approach. In regard to military operations abroad, the document affirmed the need to revise and assess each intervention carefully, avoiding the waste of resources in non-strategic areas. Finally, the *Ministerial Directive* focused on the improvements required at operational level, highlighting the crucial role played by joint commands, Special Forces, and a comprehensive approach.

Strategic Framework and Doctrine: UK and France

United Kingdom

According to the UK Ministry of Defence,[12] protecting national interests by projecting power through expeditionary interventions, 'providing security for stabilization' and 'supporting civil emergency organizations' are the main responsibilities of the armed forces. In particular, the 2012 and 2013 priorities were to bring stability to Afghanistan; undertake a wide range of military operations; fulfil defence commitments 'at home and across the world'; and continue the transformation of defence through the restructuring of the armed forces to create a simpler and more effective organization at a lower cost to the taxpayer.

If we compare the above responsibilities and priorities of the British armed forces with their Cold War posture and aims, the overall military transformation is striking. At the beginning of the 1990s, the UK Regular Forces comprised 306,000 units, while in July 2013 the personnel amounted to 160,080.[13] The transformation involved not only a significant reduction of manpower but also a 'major investment in networking of the armed forces, a fundamental shift in military mindset from doing things (often violently) to achieving effects (ideally non-violently), and a move from the large-formation force structure of the Cold War to joint modular expeditionary force packaging' (Farrell, 2008, p. 777).

Great Britain moved from territorial and alliance defence towards a new form of 'forward defence' through low-medium and high intensity crisis-management operations (Dyson, 2008).

---

12 See: https://www.gov.uk/government/organisations/ministry-of-defence/about.

13 The Royal Navy had 33,380 units, the Army 98,030, and the Royal Air Force 36,370. The Ministry of Defence has 68,770 civilian personnel. See: https://www.gov.uk/government/organisations/ministry-of-defence/about.

The radical change in the international scenario after the fall of the Berlin Wall inevitably created incentives for the UK to modify its military instrument in response to new threats and opportunities. The first review of the British armed forces had already occurred in 1990 when the Conservative Government had presented the study *Options for Change*. The document, which aimed to reorganize the UK forces after the demise of the Soviet menace, severely cut overall defence spending (Freedman, 1999). *Options for Change* advocated the amalgamation of regiments, as well as reductions in equipment, infrastructure and, mainly, manpower. Personnel were cut by 18%, especially within the Army, which was reduced from 160,000 to 120,000 units.

A further assessment of spending was released in 1994 with the *Frontline First: The Defence Costs Study*, whose purpose was to promote savings (primarily through the re-organisation of command structures) and additional manpower reduction. To this end, the *Frontline* emphasized the growing need to subcontract numerous functions to the private sector, paving the way for a massive process of outsourcing.[14]

During the 1990s UK defence devoted specific attention to joint operations and joint structures. The Permanent Joint Headquarters (PJHQ) was established in 1996. The new Chief of Joint Operations (CJO) was in charge of managing and planning all military interventions abroad (Dyson, 2008). The Joint Rapid Deployment Force (which was structured around Navy, Air and Infantry assets) was institutionalized some years later, thanks to the new *Strategic Defence Review* (SDR).[15]

In July 1998, the Labour Government presented the comprehensive review to Parliament for approval. As stressed by Cornish and Dorman (2009), the SDR was widely regarded as a model for defence planning because it involved a joint effort of 14 months combining government, research centres, academia and industry. The SDR defined the new 'security priorities in a changing world' (1998, p. 11). Expeditionary warfare and tri-service integration were crucial, and so were improving efficiency and reducing expenditure. The review also advocated reform of defence procurement policy. The Smart Procurement Initiative (SPI) was launched with the purpose of establishing a closer customer/supplier relationship between Ministry and industry, and enhancing cost control for new military equipment. In order to modernize the procurement process, the SDR introduced numerous organizational changes within the MOD, such as the Defence Procurement Agency (DPA) and the Defence Logistics Organisation (DLO).[16]

---

14 See: House of Commons Library (2008).

15 Most of the Armed Forces' helicopters were collected under a single command. In 2000 a Joint Force Harrier was established, comprising the Navy and RAF's fleet of Harrier Jump Jets.

16 The SPI was re-launched as *Smart Acquisition in 2000*, in order to develop the SPI concept.

The *Strategic Defence Review* largely focused on the concepts of international interoperability, mobility and 'operational flexibility' (1998, p. 145). Joint approaches, forces, doctrines and logistics were considered vital for contemporary British military operations abroad. Contrary to the past,[17] the Review outlined a wide range of defence missions. Peace-support operations and humanitarian interventions were defined as: 'operations other than war in support of British interests and international order and humanitarian principles' (1998, p. 151). The SDR not only called for shifts of emphasis in defence missions but also established a new role for the British military: 'defence diplomacy' (1998, p. 109) should contribute to conflict prevention and resolution through the integration of military and civilian assets. The Review portrayed the image of a 'force for good' consistent with the goals of humanitarian interventions undertaken by British troops overseas. This perspective was strictly connected with the broader framework of the UK's 'ethical foreign policy' (Strachan 2011, p. 49).

At the same time, in order to enhance power projection, the SDR stressed the need to invest in high-technology capabilities (ISTAR, C-3, etc.) in conjunction with allies (Farrell, Rynning and Terriff, 2013). As summarized by Farrell: the British military embarked on a comprehensive process of transformation 'towards a network-enabled, effects-oriented and expeditionary force posture' (2008, p. 804). During the 1990s, Great Britain had deployed troops in several multinational interventions to prevent crises, such as those in the Balkans. The SDR also recognized the risk of the 'overstretch which comes from over-commitment' (1998, p. 29). This challenge would shape the defence debate also in the following decades.

After the release of the *New Chapter of the Strategic Defence Review* (2002), which assessed the UK's defence capabilities to meet the challenges posed by global terrorism, *Delivering Security in a Changing World* was presented to Parliament in July 2004.[18] The first official document issued in the new century defined the future structure of the British military, highlighting the importance of transforming forces according to 'the characteristics of speed, precision, agility, deployability, reach and sustainability' (2004, p. 2).

The paper emphasized two decisive paths that the 'new British armed forces' should undertake: exploit the benefits of Network Enabled Capability (NEC), and develop effects-based planning and operations.[19] Some authors consider network enablement, which has been influenced by the US concept of network-centric warfare (NCW), to be the core of post-Cold War British transformation (Farrell, 2008). Others emphasize that Delivering Security in a Changing World

17 In 1993, the *Statement on Defence Estimates* still considered alliance defence to be the core objective of the armed forces.

18 The first part of the document (the so-called 'White Paper') was written in December 2003.

19 The UK doctrine moved from the concept of Effects-Base Operations (EBO) to Effects-based Approach to Operations (EBAO). See in details Chapter 5.

was deeply influenced by the US Revolution in Military Affairs, which had been intentionally set to one side by the SDR (Cornish and Dorman, 2009). Indeed, the document devoted specific attention to the acquisition of C4ISR and precision-guided munitions (PGM), as well as to Special Forces, the RAF's rapid-reaction alert capabilities, and amphibious and carrier strike task groups.

Delivering Security in a Changing World was largely based on the military experience acquired in Kosovo, Sierra Leone, and in the first years in Iraq and Afghanistan. According to the paper, expeditionary capabilities should be able to sustain three concurrent small- and medium-scale operations (rather than two in the SDR), while the most complex large-scale operations were conceived only as part of a US-led coalition. In regard to the multinational framework, although UK defence and security policy has played an important role in the development of initiatives such as the Saint-Malo Accord (1998) the Helsinki Headline Goals (1999), and the European Security Strategy (2003), it remains firmly anchored within NATO (Dyson, 2008).

The independent Haddon-Cave Review of 2009 described the 1998–2009 period as a 'deep organizational trauma' because of changes and financial cuts. According to Taylor (2012), although Labour failed to keep defence spending within proper funding levels before the financial crisis, it obtained important results in terms of management: the introduction of capability-based acquisition after 1998 (with the creation of the Equipment Capability Customer – ECC), and growing attention to long-terms needs – as exemplified by the 2007 merger of the Defence Procurement Agency and the Defence Logistics Organization to form Defence Equipment and Support (which is responsible for almost half of the defence budget).

Despite those efforts, in 2009 the defence procurement programme was still considered 'unaffordable' by the National Audit Office (Major Programmes Report), which specifically emphasized significant gaps between commitment and the resources available. The global financial crises and the high cost of operations in Iraq and in Afghanistan deeply constrained UK public spending, and did not leave the defence budget unscathed.

It is precisely the process of deficit reduction promoted by the Conservative-Liberal Democrat coalition government that is the crucial premise for the *Strategic Defence and Security Review*, SDSR (2010).[20]

The document was released together with the *National Security Strategy* (NSS) and a spending review. 'The UK was the first Western state to undertake a complete defence and security review in the age of austerity' (Cornish and Dorman, 2012, p. 213). The complex task was that of facing new multidimensional challenges to national security with reduced resources.

---

20 The review consisted of *National Security Strategy* (published on 18 October) and the *Strategic Defence and Security Review* (released on the following day). For a comprehensive analysis of the document see, among others: RUSI (2010) and Farrell, Rynning and Terriff (2013).

The SDSR proposed a more 'selective use of force' in order again to avoid the risks and costs posed by the overstretch and its legacies (2010, p. 17). A never-ending importance was attributed to investment in key technologies, as well as to flexibility and jointness.[21]

In summary, 'The Future Force' designed by the review had three broad components: the 'Deployed Forces' (those engaged in operations), the 'High Readiness Forces' (troops that can react rapidly to crises by, for example, counter-terrorism missions) and the 'Lower Readiness Forces' (those recently returned from the battlefield and those preparing to enter a period of high readiness).

In contrast to the cuts to military forces, enhanced resources (almost 0.7% of GDP) were committed to development cooperation, which was considered an integral part of UK defence diplomacy. According to Farrell, the two main components of Blair's approach were the need to face threats abroad, which is still significant, and the humanitarian goals of foreign policy.[22] Thus, also the SDSR adopted the broad comprehensive approach to defence and security policy that had initially been promoted by the Strategic Defence Review in 1998.

In terms of defence management, the *Levine Report* (2011) was the starting point for the wider Defence Transformation Agenda programme, which called for a proper balance between effective jointness and the never-ending role played by the individual services. The Report highlighted the need to adopt strong central direction so as to impose constraints preventing inevitable rivalries among services and improve cohesion through jointness and multinational interoperability. Indeed, a new Joint Command was developed after the Report.

However, the main concern was still the costs of long-term programmes (like the Eurofighter) that deeply affected overall defence spending (Cornish and Dorman, 2011). Two main features can be stressed in regard to the ways in which the UK's defence policy sought to reduce this influence. First, as had already emerged in the 1990s, significant extra funding was required to enhance capabilities on the ground. The mechanism of Urgent Operational Requirements (UORs) made it possible to provide these resources as quickly as possible. Second, the cabinet constantly defined the cuts to several military programmes proposed by the SDSR as necessary 'to fund equipment and helicopters needed for operations in Afghanistan' (Edmund, 2010).[23]

Finally, in terms of methods, the SDSR was distinguished by major innovations. Contrary to what had occurred in the past, the Review (as well as the NSS) clearly prioritized threats and interests: so that for the first time a risk assessment and management methodology is placed at the core of British security and defence

21 In 2012 a new UK Joint Expeditionary Force was established to enhance integration among the three services.

22 Interview: Farrell.

23 For a critical perspective see Farrell, Rynning and Terriff (2013).

policy (Cornish and Dorman, 2011). Moreover, new public bodies like the National Security Council (NSC) were created to support the overall review process.[24]

In summary, as well highlighted by Edmund, the UK's doctrines contain two main views of military transformation. The first is the 'power projection interpretation of military force overseas' (2010, p.388): the expeditionary approach that envisages a force structure based on new military technologies for equipment programme (such as new aircraft carriers, Eurofighter, Future Rapid Effect system, attack submarines, etc.). The second is the 'stabilization interpretation of military transformation' (2010, p.388), where the primary focus is on the maintenance of effective and sustainable ground forces (e.g., through helicopters or infantry equipment) rather than on high-intensive technology platforms.

France

The French Armed Forces that emerged from the Cold War had a legacy of relative autonomy within the Western bloc symbolized by the Gaullist choices on positioning relative to NATO and on the nuclear *force de frappe*. Also the French developed their autonomous view of insurgencies and COIN in both Algeria and Indochina, as well as less intense interventions outside Europe both during decolonization and its aftermath. The *Livre Blanc* of 1972, constituting the strategic framework for defence policy, recognized operations of this latter type as core functions of armed forces, besides territorial defence, deterrence, and the conduct of war in Europe. The same document specified the capabilities required for this set of operations, which had features similar to post-Cold War interventions, as well as assets such as foreign military bases to be used for these purposes (*Livre Blanc*, 1972, ch. 2.4).

The following *Livre Blanc*, released more than twenty years later in 1994, recognized that the defence of France no longer involved direct territorial defence, and that 'stabilization' and 'crisis prevention' would play a greater role in defining 'security' and the tasks of the armed forces (*Livre Blanc*, 1994, ch. 1.2). It maintained that an autonomous nuclear arsenal was still an essential guarantee of French 'independence and national interests' (*Livre Blanc*, 1994, p.2). This process of 'renewal' entailed a restructuring of forces, which was already undergoing, that included a reduction of personnel and equipment so as to render them more modern, efficient, and deployable (the original term was *forces projetables*). This latter concept featured prominently in the *Livre Blanc 1994*, where it was associated with a greater role of Special Forces and improved logistics to favour rapidity in the deployment of contingencies for scenarios of limited crisis (that is,

24 To be noted is that the idea of a National Security Council is one of the main proposals shaping the current debate on the content of the new Italian White Paper. See, for instance: 'Convegno nazionale sulla sicurezza internazionale e la difesa', available at: http://www.difesa.it/News/Documents/Convegno%20Nazionale%20Sic.%20e%20Difesa%20-%20Relazioni.pdf.

not directly involving national security or vital interests). Intervention in major regional conflicts was also considered to be a key task, and here the command of the air and the sea, the ability to transport and operate light armoured divisions, and to conduct joint (*interarmée*) operations.

Such tasks were reflected in the planned force structure. The Army (*Armée de terre*) should be able to deploy 120,000/130,000 soldiers, more than a half of its total personnel. Professionalization of the military should proceed, although conscription was still deemed important.

The same *Livre Blanc 1994* included scenarios for force reduction, considering the costs of professionalization as well the type of tasks envisioned, but such a scenario proved rather conservative when compared with the actual course of events. Only two years after its release, in fact, the new President Jacques Chirac suspended conscription (ended for peacetime in 2001), paving the way for an all-professional force, and in 2000 the quantitative reduction in personnel was much greater than expected.

Transformation of the Air Force (*Armée de l'Air*) and Navy (*Marine*) was less of a watershed because core functions and personnel needed less drastic changes. Here the primary focus was on shifting from tasks related to territorial defence in Europe to different types of international operations to be conducted mostly (though not exclusively) with allies and with limited (though not absent) forward logistical support. While the presence of overseas bases constituted an important asset for France's ability to deploy troops and means rapidly, the Air Force and Navy should focus on constant evolution in performing their core missions (through the use of precision munitions, new tools for surveillance and reconnaissance, and so on), and they should notably improve coordination with other services (*Livre Blanc*, 1994, chs. 5.2 and 5.3), especially in order to guarantee effective and sustainable logistics.

The decade after 9/11 was a period of intense military activism that in France concerned the production of military doctrine as well. President Nicholas Sarkozy in particular provided a strong push re-entering NATO Command Structure and strengthening relationships with the US. France's inaction in Iraq contrasted with its strong presence in Afghanistan and, most recently, with its driving role in interventions in Ivory Coast (2011), Libya (2011) and Mali (2013, under President Hollande), which signal both a political desire for strong international presence and, most importantly, the existence of armed forces able to guarantee the deployment and actual use of military force in relative autonomy. This must be also understood as a 'cognitive' autonomy – as rapid inspection of the vast amount of available doctrinal production easily shows. It is difficult to make a coherent summary of such a large *corpus*; nevertheless, it is possible to identify a few headlines, the focal points of the transformation of French armed forces.

The *Livre Blanc* released in 2008 presented the overall framework within which specific doctrines were developed. Besides territorial defence and dissuasion, three major missions were contribution to international peace and stability operations, intervention in a major conflict abroad, and autonomous capacity of reaction.[25] Major intervention would require a force of 30,000 troops, 70 airplanes and a naval force able to project force (*Livre Blanc,* 2008, pp. 221–2), and it was envisaged within a multinational framework. The autonomous capacity was based on the ability rapidly to deploy 5000 troops on the ground together with aerial and naval power projection and support (*Livre Blanc*, 2008, p. 222).

In terms of numbers, the Army's total personnel amounted to 131,000 people, 88,000 of which should be operational. The design of the eight combined arms brigades included assets for a large set of missions, including relatively large-scale conventional operations with heavy tanks and close-air support. The Navy should employ 44,000 people and have three main functions: dissuasion to nuclear-armed submarines, strategic logistical support, and power projection even in distant theatres (7,000/8,000 km). The Air Force, composed of 50,000 people, would set up a Joint Forces Air Component Command (JFACC). This would be able rapidly to deploy an emergency response force of 1500 troops, while guaranteeing strategic and tactical mobility and the use of about 70 airplanes (out of the total of 300 available) in a major conflict.[26] Joint forces structures would play an important role in ensuring the coordination of command and planning in joint operations, information warfare and special operations.

Other documents have reshaped French doctrine. They rest on three cornerstones. First, according to the *Concept d'Emploi des Forces* (Ministère de la Défense, 2010) released in 2010 (and coherently with the NATO *Strategic Concept* of the same year) the new strategic environment raised threats of a new kind, multi-dimensional and complex, which were to be faced through multilateral action, on the one hand, and through a set of capabilities including integrated political and military action, on the other. This integration was necessary both because the nature of new missions (COIN for instance) required armed forces to take a non-military approach to certain problems, and because technological transformation in the field of communications was (sometimes) making decisions at the tactical level politically relevant (Ministère de la Défense, 2010, pp. 21–7).

Second, the nature of military action was such that operations were necessarily joint-forces operations (Ministère de la Défense, 2010, pp. 27–8). Interoperability, defined as the ability to intervene interdependently with other countries, was a key asset (Ministère de la Défense, 2010, pp. 34–5). The United States was explicitly mentioned as the ally to take as a reference (Ministère de la Défense, 2010, p. 34).

Third, power projection was a fundamental asset (Ministère de la Défense, 2010, p. 36). This meant that armed forces should be readily deployable and able to be used in a diversified set of strategic and operational environments. These included far spots thus strengthening the need for coordinated large-scale

25 Interestingly, the other (sixth, but first in the list) mission to be performed by armed forces was to 'ensure the knowledge of risks and threats' (*Livre Blanc 2008*, p. 221).

26 Deterrence, surveillance of national territory and, relatedly, of space were other major missions of the Air Force.

logistics. It also meant that French armed forces should be able to sustain the prolonged effort of an expeditionary force through, for instance, the appropriate use of operational reserves.

Despite all the hype for a document released after two important missions – the interventions in Libya and Mali, the latter conducted almost autonomously, and in the midst of an economic crisis – the new *Livre Blanc* released in 2013 exhibits a fairly limited amount of novelty, at least in the narrative. The document recognizes substantial continuity in the identification of increasingly heterogeneous threats requiring constant redefinition of what constitutes 'national security'. Likewise, the *Livre Blanc 2013* also recognizes the substantial stability of the French areas of interest: the Mediterranean, Africa to the Equator, the Persian Gulf, and the Indian Ocean. It acknowledges that France's capacity to cope with these challenges requires that, while measures to improve efficiency of spending are possible, the country's overall economic effort must remain at least constant in absolute terms in the period 2014–2025 (the horizon of the document).

This effort is intended to ensure the following capabilities for 2025:

- An immediate reaction force (so-called *force interarmées de réaction immediate, FIRI*);
- The ability to deal as the main actor with two or three distinct missions of crisis management, through forces equivalent to a joint brigade of 6,000/7,000 men with a number of special sea and air forces for the mission's proper fulfilment;
- The ability to deal with more intense conflict, even in a coalition to be guided by France through two joint brigades of 15,000 men composed of special ground, sea and air forces.

Given these requirements, the force structure should be organized as follows:

- Land Forces: seven joint brigades with 66,000 men, with two 'high readiness' brigades able to deal with opponents equipped with heavy weapons. Specifically, the French forces should have (approximately) 200 heavy tanks, 250 medium tanks, 2,700 multi-role armoured vehicles, 140 attack and recognition helicopters, 115 transport helicopters, and 30 tactical drones;
- Naval Forces: 4 nuclear submarines with ballistic missiles, 6 attack submarines, 1 aircraft carrier, 21 frigates (15 'First Rank' and 6 'Surveillance'), 15 patrollers, as well as other means of logistical support and force protection (like minesweepers);
- Air Forces: about 225 combat aircrafts (including the Navy's aerial forces), 50 aircrafts for tactical transport, 7 aircrafts for aerial surveillance, 12 aircrafts for aerial refuelling, 12 drones for reconnaissance and surveillance missions in theatres, 8 land-to-air medium-range missile systems.

This structure closely mirrors the one envisaged in 2008. Notwithstanding the recognition of the importance of maintaining a force structure coherent with the ambitious missions set for French forces, and of planning expenditures accordingly, there is an actual scaling down of forces in comparison with the *Livre Blanc* 2008. A closer look at budget structures over time sheds more light on the drivers of these dynamics.

The Italian Defence Budget from 2001 in Comparative Perspective

General Trends

Figure 3.1 shows the evolution in time (2000–2011) of the public expenditures allocated to Defence. More precisely, as defence budget in Italy includes resources for *Carabinieri* involved in domestic policing, here we refer to the *Funzione Difesa* (Defence Function) section of the budget.[27] The decline of resources brought that budget to below 1% of the GDP (for 2011, that is 14.36 billion euros), well below the threshold of 2% pledged as a NATO member.

The figure would increase if one adds the separate budget devoted to international missions (which is not part of the defence budget, but constitutes a different entry) and the funds allocated to defence-related investments by other ministries. In particular, the Ministry for Economic Development (MiSE), has funded and funds major programmes involving Italian industries. As for the extra-budget devoted for missions (approved yearly or every six months), the average yearly expenditures in the 2000–2011 time-span has been beyond 1 billion euros, with a large percentage (almost 70% in the most recent years) related to Afghanistan (Forte and Marrone, 2012). A precise estimate of the contribution of MiSE is more complicated, but the Eurofighter airplane, the FREMM frigates and other aerial and space systems have been partly (or in cases such as the FREMM largely) funded by contribution that in recent years topped 2 billion euros. Summing these contributions would bring current expenditures above 1% of the GDP.

Needless to say, such restraint would have a huge impact on force transformation no matter the way it is then allocated, as it does not provide (much) room for experiments to go beyond mere subsistence. While this is, quite naturally, the mantra of the military-industrial complex, it remains nonetheless a quite realistic painting of the situation. No matter how doctrinal preferences evolve, the equilibria that such budget allows are always relatively lower than those by other major European countries.

---

27 Data are taken from the Ministry of Defence budgets published each year from 2000 to 2012. The *Nota Aggiuntiva* (the report attached to the yearly budget) is the source of detailed information on allocation of resources and programmes. It can be found on the website: http://www.difesa.it/Approfondimenti/Nota-aggiuntiva/Pagine/default.aspx.

In fact, France and Great Britain show a quite different picture considering the budget, which is not affected by the Italian disequilibrium because the costs for personnel are similar to the other items. In addition, contrary to Italy, the percentage of resources attributed to personnel decreased since 2001 in the case of France, while it has remained stable in the British budget.

Frequent cuts of manpower resources (in the UK) and constant investments in operations and equipment (in France) have surely played a relevant role. The legacies of the Cold War have been significant for the three countries, which had a very different military structure during the bipolar era, but the effect of different reforms approved by Paris and London in the last two decades have certainly contributed to shape a more balanced budget.

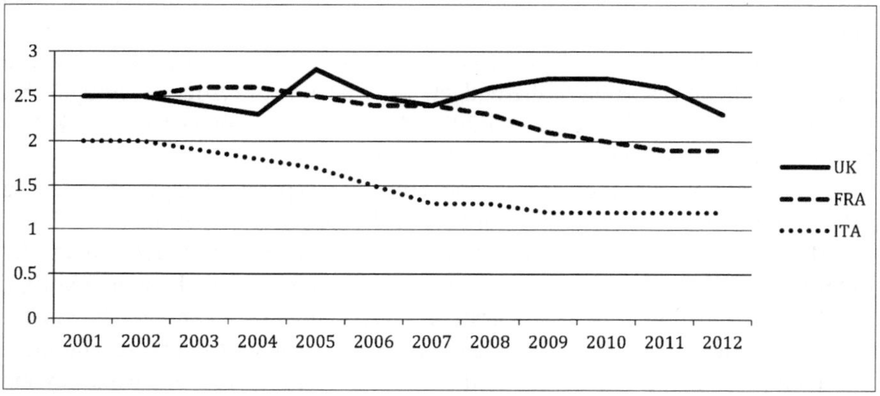

Figure 3.1 Defence expenditures over Gross Domestic Product
Source: NATO Information on Defence Expenditures, 2001-2013 (Constant prices).
Note: All data are elaborated by the authors.

Defence Spending by Service

Figure 3.2a shows the allocation of resources among different services. Army (*Terrestre*) is consistently the major recipient of resources, largely due to personnel costs. Navy (*Marittima*) remained relatively stable, while Air Force (*Aeronautica*) suffered the major cuts, while Joint activities (*Interforze*) have risen until recently. It must be noticed, though such a rise is largely due to expenditures for the new F-35 Joint Strike Fighter (JSF) being located under Joint activities rather than Air Force. On one hand, this contributes to explain the cuts to the latter's budget, and on the other it helps to understand the lack of open confrontation over them: the Air Force will directly manage about the three-quarters of the JSF's intake, also being the only aerial component among allies that will also directly manage the vertical landing version of the plane (F-35B). The planned reduction (as of 2013) of the total number of JSF that Italy will buy (from the planned 131 to the current 90) might actually create further inter-service rivalry, as the number of F-35B available to each service will be so limited (the allocation is under discussion, but it should be of 15 units/each) to make it difficult to operate autonomously. Figure 3.2a breaks down investments according to services.

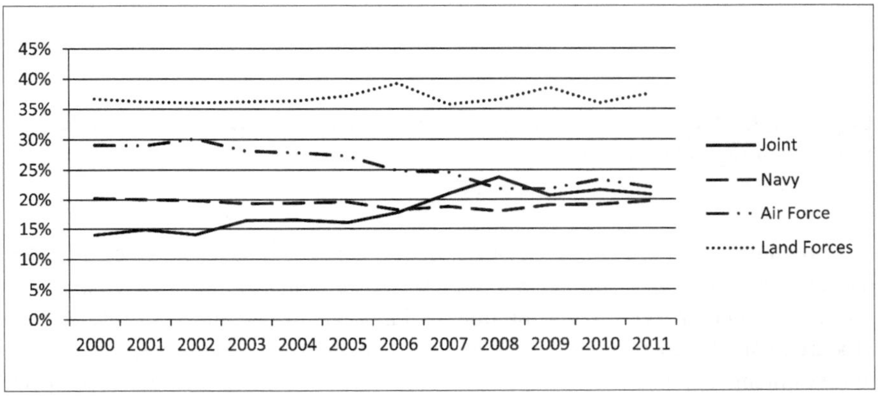

Figure 3.2a Expenses by services over total defence expenditures – Italy
Source: Italian Ministry of Defence budgets 2001–2012.

Figures 3.2b and 3.2c shows the pattern of spending by service respectively in the UK and France. Data released are organized differently in the UK, which explains the different structure of the graph and the lack of information about Joint services. Still, also Great Britain established in 1996 a Permanent Joint Headquarters (PJHQ) that manages of all military interventions abroad.[28]

28 The report of the distribution of resources among services changed in 2003. Instead of having an aggregate data for Army Navy and Air Force, the UK's defence budget distinguish among Commander-in-Chief Fleet – Naval Home Command and Defence Equipment & Support 2nd Sea Lord; Commander in Chief-Air Command and RAF training command; Commander-in-Chief Land Command.

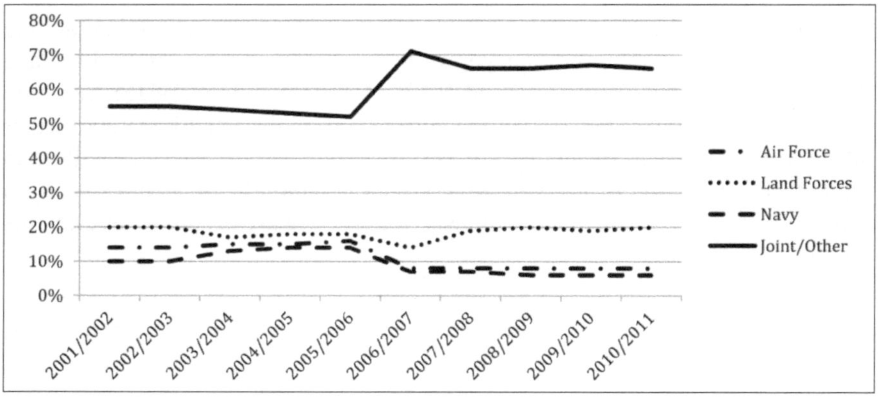

Figure 3.2b Expenses by services over total defence expenditures – UK[29]
Source: Ministry of Defence; Defence Analytical Services and Advice 2001–2012.

A re-organization of the French structure of Defence that took place in 2006 had already given to Common Services linked to Joint Commands (*Commande Interarmées*) control of a larger number of functions (especially with reference to procurement), which explains the extent of change observed.[30] With special reference to investments in equipment, spending over *services communes* (joint) moved from substantial parity with other services at the beginning of the century to account for beyond the 75% threshold in recent years (Ministère de la Défense, 2012).

29 The UK defence budget presents notable differences in terms of structure. Two points must be noticed. First, from 2001, the UK moved from a 'cash based accounting system' to a 'resource based system'. Different items and sections changed. For this reason, we can compare only the expenses related to 'personnel' and 'equipment' while the definition attributed to 'operations' varies across time, making impossible to report aggregate data. Second, aggregations by service are not provided by official statistics. Here, we aggregated as follows: Navy (Commander-in-Chief Navy Command; 2nd Sea Lord/Commander-in-Chief Naval Home Command), Air Force (Air Officer Commanding-in-Chief RAF Personnel and Training Command; Air Officer Commanding-in-Chief RAF Strike Command; Commander-in-Chief Air Command). Land Forces Commander-in-Chief Land Forces); Joint/Other (Chief of Joint Operations, Chief Defence Logistics, Central Defence Estates, Defence Procurement Agency, Science Innovation Technology Administration, Defence Equipment & Support, Adjutant General – Personnel and Training Command). The data considered are relative to the Department Expenditure Limits (DEL), limitedly to the entry Resources (thus excluding Capital). For full reference to terms, see Ministry of Defence, 2012.

30 More specifically, the BOP (Budget Operational Programme) of Programme 178 of the Ministry of Defence deals with distribution of resources among services (including common services) and the BOP of Programme146 with the allocation of expenses devoted to equipment. The 2006 Programme 178 can be found at: http://www.performance-publique.budget.gouv.fr.

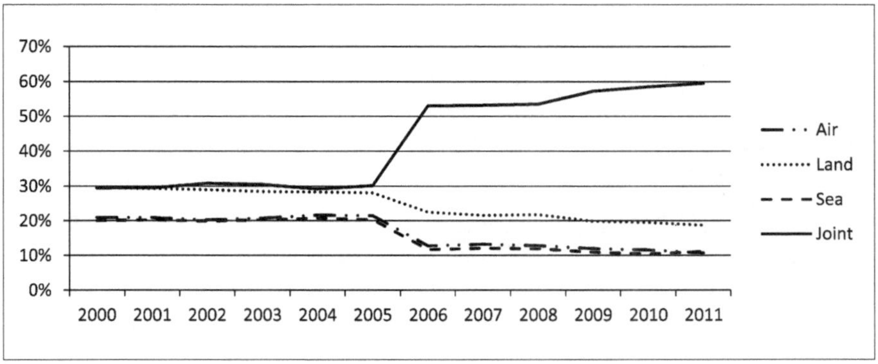

Figure 3.2c Expenses by services over total defence expenditures – France
Source: *Annuaires Statistiques de la Défense* 2003–2012.

Expenditures by Function

A glance at allocation of resources among different budget entries (personnel, operations, investments) allows examining the situation more in detail (Figure 3.3a). As it can be seen, expenditures cover mostly personnel, accounting for an average of the 58% of the budget for the entire period (and for 64% in the last 5 years). The Law 66 of 2010 that confirmed the total number of personnel established in the laws on the *New Defence Model* from the Nineties and thus prolonged the heavy burden of personnel-related cost (Mason, 2013). Investments never accounted for more than 30% of the budget, while operations have been steadily declining (and received around 10% of the budget in 2011). The same sources openly admit that an optimal budget would entail a substantially symmetric division of resources. The Report that is attached yearly to the budget (*Nota Aggiuntiva*) noticed the lack of sustainability of the existing ratio between planned personnel and effective personnel as well as between personnel and operations and maintenance expenditures (Ministero della Difesa, 2010). The importance of this asymmetry, as well as its persistence (with changes in components) over time, can hardly be overstated. Lack of money channelled in operations and maintenance vastly affects the ability to rapidly deploy and effectively employ the available forces.[31] Funding for training and education of officers also suffered.

As it has been noticed in the Army Report, allocation to these functions in 2011 is slightly more than 20% of what it had been in 2003 (SMD, 2012b). As stressed above, extra-budget was specifically devoted to missions and industrial investments constituted at least a partial remedy as in recent years it funded respectively essential operational requirements (with deployments representing a key moment for 'testing' as well as actual use) and innovation in equipment.

31 Osservatorio di Politica Internazionale (2010).

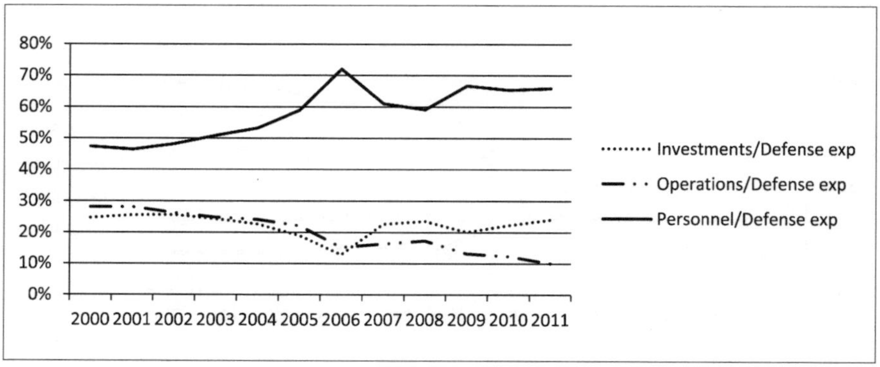

Figure 3.3a Expenditures by major entry over total defence expenditures – Italy

Source: Italian Ministry of Defence budgets 2001–2012.

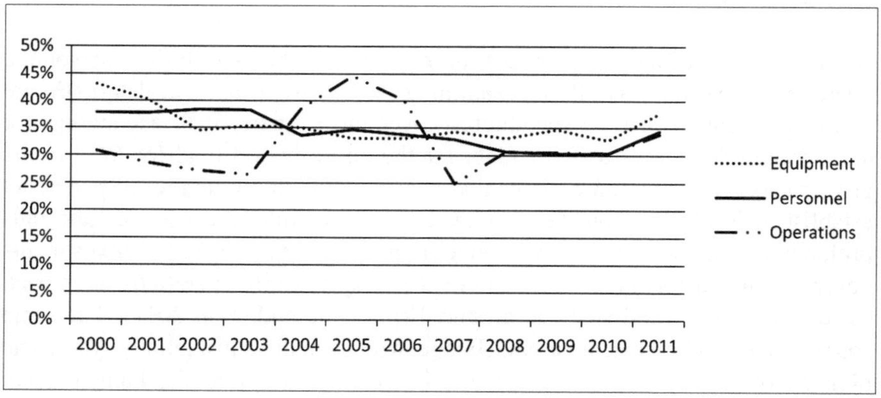

Figure 3.3b Expenditures by major entry over total defence expenditures – UK

Source: Ministry of Defence; Defence Analytical Services and Advice 2001–2012.

Figures 3.3b and 3.3c show the pattern of spending by major entry in the UK and France. Differences are quite evident both in terms of the weight of personnel expenditures (significantly lower) and of the equipment/investment expenditures (significantly higher).

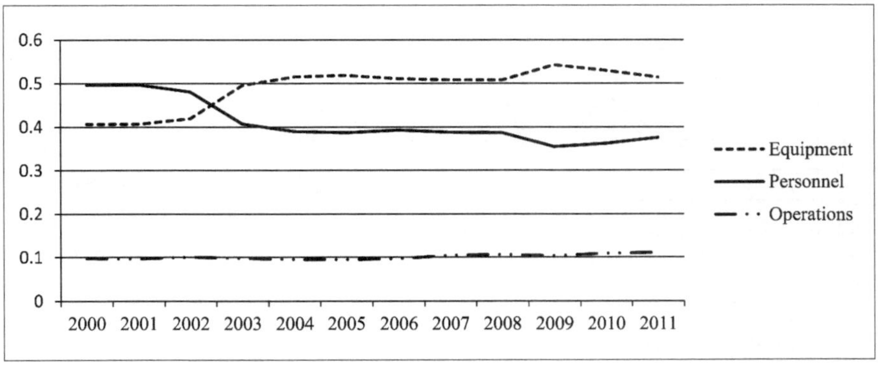

Figure 3.3c Expenditures by major entry over total defence expenditures – France

Source: *Annuaire Statistique de la Défense*, 2003–2012.

Finally, Figure 3.4 breaks down investments of the Italian armed forces in major programmes (MP).[32] The acquisition of the NH-90, of the Eurofighter and of the F-35 and the new aircraft carrier *Cavour* (designed to harbour the F-35B) as well as the new class of submarines U-212 (a joint project with Germany) and the Orizzonte-class of Frigates have drained and still drain some much of the investment part of the budget, and they will in the future consume a large part of the already underfund budget devoted to operations. These programmes were decided well before 2001, and are fairly unrelated to missions undertaken in the past decade. This excluded of the NH-90 (dispatched to Afghanistan quite recently) and the tactical vehicles of the Army (*Lince*) as well as the Eurofighter (and the enhanced Tornado, another relevant entry of expenditure) for the intervention in Libya. Investments in airlift capabilities, often advocated as a key component of modern forces, have been fairly limited and Italy recurred to lease other countries' planes for heavy-lift. In the Appendix, more information on major programmes undertaken is provided.

External Factors and Actors: The Role of NATO and EU

The evolution of national doctrines and force structures comes about in a highly institutionalized context, of which NATO and the EU represent two cornerstones. Although they do not directly form a key component of defence policy, or the budget, they nonetheless exert significant influences on doctrine and planning.

32 By major programmes here we defined single budget entries that accounted for more than 10% of the total investment-related budget.

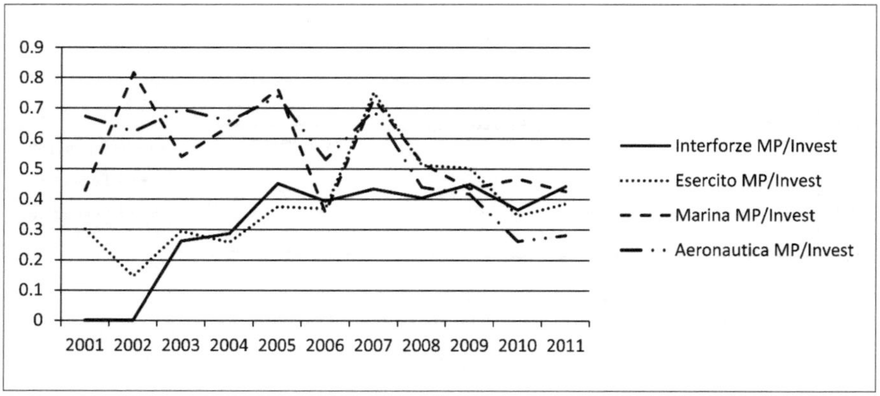

Figure 3.4 Expenditures in major programmes
Source: Italian Ministry of Defence budgets and *Nota Aggiuntiva*, 2001–2012

In terms of force structure, one feature should be stressed. Italy's post-Cold War commitment to international missions is strictly related to participation in multilateral institutions. The Ministry of Defence distinguishes operations by the Italian forces into five types: NATO, EU, UN, multinational, such as the MFO (Multinational Force and Observers) in Egypt and bilateral, such as DIE – *Delegazione Italiana Esperti* (Italian Delegation of Experts) in Albania.[33] The overall Italian contribution to EU operations is not comparable, in terms of units provided, to NATO and UN interventions (Coticchia, 2014b). In fact, with few exceptions (Operation Atalanta, the anti-piracy mission in the Horn of Africa and the EUFOR Althea operation in Bosnia Herzegovina), the number of personnel employed in the ESDP/CSDP interventions has been invariably low. Despite growing engagement in EU operations since 2003 (in terms of number of missions undertaken over time), the degree of troops strength has remained limited (Marchi Balossi Restelli, 2011). Moreover, notwithstanding the constant rhetoric on the need to develop a truly integrated European Force, concrete steps in this direction have been few.[34]

On the other hand, the post-2001 Italian contribution to UN operations has been important, mainly thanks to the UNIFIL mission in Lebanon, where Italy has regularly deployed a considerable number of troops since 2006 (Trombetta, 2007). More broadly, the large majority of the post-WWII military interventions abroad have been UN operations.[35] Hence UN PKOs' rules, procedures and approaches

33 For an overview, see: http://www.difesa.it/OperazioniMilitari/op._intern_corso/.

34 See, for instance, the parliamentary debate of the Defence Commission preliminary to the European Council of December 2013 (*Commissione Difesa*, 2014).

35 Since the end of WWII, Italy has carried out 34 UN operations out of 90 missions undertaken within a multilateral framework. See: http://documenti.camera.it/leg16/dossier/Testi/di0238.htm.

(especially in the last two decades) have played a major role for the Italian armed forces. In fact, the preferred multilateral frameworks the Italian troops have contributed to improve some key tools for contemporary peacekeeping operations, such as coordination among civil and military actors or the training of local forces (de Guttry, Sommario, and Zhu, 2014).

However, mainly because of the preeminent functions of KFOR (Kosovo), Allied Shield (Horn of Africa), Unified Protector (Libya) and, above all, ISAF (Afghanistan), NATO's missions have affected the transformation of Italian forces more than any other multilateral framework. New tasks, new combat scenarios, different requirements in terms of equipment, interoperability and attitudes, are some of the foremost novelties that each NATO member has had to deal with in recent years.

The next chapter focuses specifically on the dimension of deployment, considering the ways in which needs on the ground have interacted with doctrine to affect the transformation of the Italian armed forces. But on looking at the massive amount of resources employed in those NATO operations, it is reasonable to conceive interventions like ISAF (the most costly, difficult and dangerous operation undertaken since WWII) as crucial turning points for the force structure of Italian defence.

Overall, multilateralism should be interpreted not only as the main guideline of Italian foreign policy, as clearly demonstrated by the literature (Bonvicini et al. 2011; Ratti, 2011), but also as the principal tactical and strategic framework within which Italian forces have been developed and transformed in the past two decades.

Before assessing more in full deployments, both in in terms of its coherence with doctrine and of its influence, we should emphasize two significant aspects relative to the role of doctrines drawn by multilateral organizations.

The first aspect is the shared notion of the multidimensional nature of threats. In fact, all the post-2001 doctrines crafted at national, European and NATO level focus primarily on the non-military dimension of security challenges. While concern about military attacks on national frontiers by 'traditional' enemies such as states or alliances of states (i.e. the Warsaw Pact) imbued the Cold War documents (the White Papers of 1977 and 1985), the post-bipolar doctrine has increasingly paid attention to threats stemming from more or less geographically proximate weak, failing or failed states: organized crime, piracy, cyber-attacks and terrorism (Buzan and Hansen, 2010).

The *European Security Strategy: A Secure Europe in a Better World* (ESS, 2003), identifies five main menaces for the EU: terrorism, the proliferation of weapons of mass destruction, regional conflicts with an international impact, organized crime, and failed states. The updated ESS version of 2008 (*The Report on the Implementation of the ESS*) confirms this multidimensional nature of security.

On the other hand, the most recent NATO Strategic Concept was adopted in 2010. Along with proliferation of weapons of mass destruction and ballistic missiles, all the other challenges have a non-military nature, such as 'dependence on foreign energy suppliers' or 'environmental and resources constraints'. The doctrine, which still defines NATO as primarily a 'nuclear alliance', emphasizes the need to develop capabilities for smaller crisis-response operations. Collective defence, crisis management and cooperative security are conceived as the three main tasks of the Alliance. As stated by the document, NATO will endeavour to 'prevent and manage crises, stabilize post conflict situations and support reconstruction' (2010, par.20). Therefore, as shared by the Italian official documents, a wide range of activities should be undertaken in current conflicts, which are characterized by new actors and menaces. Moreover, the Concept remarks that the 'comprehensive political, civilian and military approach is necessary for effective crisis management' (NATO 2010, par. 21).

The second aspect to be stressed is the influence exerted by the so-called comprehensive approach, which has been developed at NATO level and then gradually considered by national doctrines to be a cornerstone of their strategic view on contemporary warfare. The 'comprehensive approach' has largely shaped the British doctrine in recent years. As illustrated above, the multi-dimensional and broader perspective of security had already been promoted by the *Strategic Defence Review* of 1998. At the same time, the National Security Strategies of 2008, 2009 and 2010 still focused on the importance of non-military threats for the UK's national security. The *JDP 3-40 – Security and Stabilisation: The military contribution* (2009) is one of the most recent attempts better to articulate a holistic approach to current military intervention. As stated by Griffin, doctrinal developments have tried to 'rationalize intervention, combining elements of counterinsurgency, counterterrorism, peace support and state-building in a way that reflects their experiences in these seemingly intractable conflicts' (2011, p. 317). In a broader context of strategic reflection by NATO allies on ways to win in Iraq and Afghanistan, expectations regarding a supposed British excellence in counterinsurgency have clashed with weaknesses in the British campaigns (Egnell, 2009; Ucko, 2010). The UK deployment in Basra and Helmand will be examined in the next chapter, but worth noting from a doctrinal perspective is the problematic 'attempt to differentiate classical British counterinsurgency from contemporary stabilization' (Griffin, 2011, p. 324). For instance, the British COIN doctrine (*Army Field Manual* – AFM-10 2009) relegates counterinsurgency to the tactical level, distinguishing it from peace support and stabilization, which is considered by Farrell to be the 'key concept' for the future of British defence.[36]

More than the *Strategic Concept* of 2010, the 'unprecedented' level of integration within the NATO framework (Edmund, 2010, p. 383) is the significant source of influence for the development of a national perspective on the controversial 'comprehensive approach'. The influence of NATO extends beyond sharing a multidimensional view on current operations. In fact, interoperability with NATO is one of the core components of the UK doctrine – as well illustrated by the White Paper of 2003. Moreover, as stressed by Taylor, the NATO's 'Pooling&Sharing' may be the best alternative way to 'maintain knowledge and

36 Interview: Farrell.

core capabilities' in the case of cancellation of military programmes, as it has occurred especially since 2010.[37]

France's more complex relationship with NATO, as exemplified by its withdrawal from the Alliance's Integrated Command from 1966 to 2009, does not mean that NATO has exerted little influence over its doctrine. More than other sectors, nuclear forces maintain more autonomy, and France does not currently participate in the NPG (Nuclear Planning Group) devoted to devising NATO's policies on the issue. Such autonomy, which partly remains for conventional forces as well (in peacetime, French forces are not placed under permanent NATO command, French *Livre Blanc* 2008), should not mask the fact that France operated in close coordination with NATO well before it re-entered the Integrated Command. France participated in missions in Bosnia for more than ten years (through IFOR and SFOR), in the joint airstrikes in Kosovo, and it was heavily involved in ISAF in Afghanistan.

In this context, France has been moving towards further integration in defence systems to enhance interoperability, in particular in the fields of command and control and in missile defence (although it has been critical of the consequences of relying on missile defence as opposed to deterrence: Vedrine, 2012). The *Vedrine Report* of 2012 assessed the first years of France's renewed participation in NATO, and stressed that full participation in the Alliance is inevitable. However, while this requires further improvement in common standards, national sovereignty on key issues should be maintained. This approach contributes to explaining the limited adoption of measures (including those contained in the concept of 'Smart Defence' which arose from the 2010 NATO Summit in Chicago) that call for specialization and complementarity, thus reducing national autonomy in certain fields. Similarly, French military doctrine also stresses interoperability with NATO Allies as a key element of the *Concept d'Emploi des Forces* (CICDE, 2013), while it also emphasizes that 'future engagements may lead our Armed Forces to operate with local or contingency forces or entities that might operate according to standards different from our own' (CICDE, 2013, par. 35). Whence derives the need for French forces to identify other potential partners and build shared capabilities with them as well. In more operational terms, as symbolized by the *Doctrine d'Emploi des Forces* (CICDE 2014), the link with NATO is more visible, and it is strictly connected with NATO's *Allied Joint Doctrine* (the reference document is AJP-01 of 2010).[38]

The comprehensive approach had been often invoked in French documents as essential for managing the complexity of current crises, and integrating military and civilian capabilities (*Livre Blanc*, 2008, p.58). Yet it is difficult to go beyond this idea of convergence and think that NATO has shaped France's doctrine directly. In fact, it is possible to find ideas related to the comprehensive approach in concepts such as 'global manoeuvre' present in French strategic thinking in the

37 Interview: Taylor.

38 Interview: LTC Garnier, Armée de Terre.

twentieth century (Wendling, 2010). As will be noted in Chapter 5, this similarity in the development of concepts does not always lead automatically to the adoption of similar practices in the field (as in the case of Provincial Reconstruction Teams – PRT).

Summary: Coherence and Convergence in Defence Policy

The shared elements of post-2001 military doctrine in Italy, UK, and France can be summarized as follows. First, armed forces should be expeditionary. Power projection capabilities aim to deal with the multidimensional threats posed by regional and global instability. For this reason, naval and aerial mobility, operational readiness, logistical support and integration with multinational systems and equipment are vital. According to the doctrine, only joint coordination, multinational and interdisciplinary interoperability can guarantee effectiveness and sustainability. Second, while crisis management and peace support operations should be extremely flexible, combining force, diplomacy and intelligence (mainly through civil-military cooperation), combat missions (which have been clearly conceived only after the Chief of the Italian Defence Staff Strategic Concept 2005) require precision firepower, deployment of Special Forces, and amphibious projection capabilities. In contemporary warfare, training and cultural preparation are essential. Third, armed forces should primarily develop C4-ISTAR and net-centric capabilities, constantly improving and acquiring new technologies (especially in the fields of unmanned vehicles, electronic warfare, maritime surveillance, satellite communication). The survival and protection of the forces has always been considered a priority. Finally, it is worth noting that all documents call for a more balanced budget, pressing for the harmonization of costs, modernization, rationalization, and (above all) additional resources.

Italy, France and Great Britain thus share a similar view of the strategic context in which they operate, although the scale of their operations is definitely not identical. Other dimensions exhibit elements of distinction, often in terms of degree rather than quality. First, the allocation of resources seems to be very path dependent. This is of course a common and natural feature of defence planning and budgets, but the interplay of two factors makes it very problematic. One is the ever-present limited amount of resources, which leaves no space for innovations that have not been anticipated a long time previously. The other is the long-term nature of the commitments already assumed, because the large procurement programmes mentioned above tie the hands of both political and military decision-makers. In the case of Italy this has led to partial 'exploration' of potential requirements – as shown for instance by the budget for research and development (an important, if not complete, indicator of explorations), which is close to zero. Admiral Di Paola, who would subsequently become Chairman of the Joint Chiefs, noted in 2006 that the 'scant' resources devoted to the armed forces had induced the Italian Army to specialize in peacekeeping and peace-building operations, thereby inhibiting different paths of development.

This type of constraint generates disconnection across levels, and especially a rupture between the politico-strategic and the operational and tactical levels that will be explored more thoroughly in Chapter 4. There may also be a paradoxical unintended positive consequence of this situation: Military organizations will not be able to over-react to emergencies and adopt changes that – while making sense in the short run – might risk endangering sustainable innovation in the medium and long term. In other words, the relatively limited influence of lessons from COIN, especially in Afghanistan, on budgetary allocation may actually have prevented the diversion of resources to activities of a type that do not figure prominently in current planning (contrary to before 2001). The fact that major programmes were decided in the 1990s (or earlier) has made a change de facto impossible.

For sure, there are downsides. One is that a lack of flexibility means a relative inability to cope with emerging crises clearly signalled by problems encountered by the armed forces. For instance, although the aerial campaign in Libya was marked by success, it showed the inherent limitations in terms of quality and quantity of the resources available, because the intervention required the assets of the US in early phases and, exhausted fuel, resources for maintenance, and smart ammunitions (Mason, 2013)

All European countries faced significant challenges during the Libyan war. They illustrated the consequences of the recent decisions imposed by financial constraints. Italy had to withdraw its carrier *Garibaldi* in the midst of the conflict because of budgetary pressures.[39] The RAF's operations were considered under threat at the end of June 2011 if the campaign had continued. According to Cornish and Dorman: 'If the Libyan operation had lasted for longer than six months, then the Ministry of Defence's planning assumptions would have been outstripped' (2012, p. 217). The scrapping of the Harrier programme had affected air power capabilities. Also France withdrew its *Charles de Gaulle* carrier for maintenance reasons.

While the Unified Protector operation once again revealed the importance of the NATO framework, the degree of military cooperation between France and Britain was limited.[40] Nevertheless, as a report of the Royal United Service Institute noted, one of the legacies of what has been perceived as a successful military undertaking by France and UK may be a strengthening of (political and strategic) bilateral relations as it becomes clearer that the two countries are unique in Europe in their ability – and willingness – to conduct combat operations (Johnson and Mueen, 2012). Overall, the Libyan campaign remained 'heavily dependent on the US to provide ISTAR capabilities. In particular, European nations had to rely on

39 See Defence News (2011).

40 Interview: Chalmers. A slightly more optimistic view of the cooperation, notwithstanding recognition of coordination problems in key phases of the campaign (such as the beginning of air strikes) can be found in Cameron (2012).

US joint surveillance target attack radar systems (JSTARS) and airborne warning and control systems (AWACS) aircraft' (Hallams and Schreer, 2012, p. 323).

In general, in terms of co-evolution between budget and doctrine, the Italian budget seemed to have co-evolved with a stress on joint activities but was less focused on PKO than doctrine. In particular, the rising costs of specific military assets such as airplanes and ships provoked some inter-force disputes between the proponents of the need of expensive 'conventional' and hi-tech equipment and those who pointed out that the prevalent activities of the Army in irregular warfare require different choices and tools (Fraticelli, 2006; Mini, 2008). The responsiveness to changes occurring both in doctrine and the field is actually constrained by stickiness of existing commitments (the Appendix provides data on the acquisitions of the procurement process related to major programmes). The slow process of troop number reduction also points to resistances to budget cuts by the Army (the most manpower-intensive service) and broader political worries about limiting overall employment levels.

Also in the UK the debate was mainly focused on treasury-imposed cuts and the resulting 'inter-services rivalry and tribalism' (Cornish and Dorman, 2011, p.337). The then Chief of the General Staff of the British armed forces, Sir Richard Dannat, identified over-committing and under-resourcing as the main problems for the UK defence in last two decades (Edmund, 2010). As illustrated by analysis of the doctrine, post-Cold War British defence struggled to adapt to growing challenges, firstly through an overall reduction of forces: from 1989 to 2009 the UK reduced half of its major vessels and 41% of service personnel (Chalmers, 2010).

The SDSR reduced the national defence budget by 7.5% (Taylor, 2012). The review notably cut across the Armed Forces, reducing personnel by 17,000 in all three services and curtailing the UK military presence abroad. 'Bringing the defence budget back to balance is a vital part of how we tackle the deficit and protect this country's national security' (SDSR 2010, p. 3). As reported by Edmund (2010), the government clearly identified low priority areas, such as fast jets or the civilian staff at the Ministry of Defence. The consequences of SDSR have been manifold: the closure of several bases, the cancellation of numerous military programmes (e.g. the Nimrod MR4 maritime patrol aircraft), the withdrawal from service of Harriers, and the shrinkage of the Tornado GR4 force. The RAF is moving from fast jets to providing services and transportation. Moreover, several units, such as the Type 22 frigates, have been retired from service without replacement. Above all, the UK will have no fully operational aircraft carrier until 2020.

The UK official strategic documents have tried to prepare the forces for the full spread of capabilities, facing at the same time increasing unit costs for new equipment because of qualitative improvement. The expeditionary approach requires expensive rapid and joint interventions at long distance, where the cost of force protection has considerably increased in the new century.

In a context of budget shrinkage, growing importance has been attributed to jointness – as well illustrated by the above-mentioned *Levine Report* of 2011. Overall, numerous objectives have been achieved, such as the Joint Command, which guides military operations and is also responsible for the elaboration of doctrine. However, some obstacles still remain, in both cultural and organizational terms. For instance, career progression is related to the single service, and additional incentives to improve jointness (especially at medium and low level) are still required.

Transformation in France is driven by the double objective of maintaining autonomy while recognizing the need to strengthen interoperability within NATO. Of course, this approach does not always guarantee the maximization of efficiency because some redundancy is inevitable. At a time of intense use and budget limitations, this creates notable problems. Analyses of budget transformation in France highlight some critical issues. One of them is that, while the defence budget may change even abruptly in short periods of time, some of the costs (in particular those of personnel) are relatively sticky (Foucault, 2012). This implies that expenditures on equipment are those most subject to change. Combining this trend with the increase in the unit-level costs of military equipment (in France as elsewhere) since the 1990s, the result is that modernizing equipment becomes increasingly difficult. Relatedly, the Military Programme Law (MPL) that establishes the amount of resources to be devoted to defence spending has often been more a declaration of intent than a clear directive, with adjustments being made across the six years of the MPL, almost always at the expense of equipment modernization (Foucault, 2012). Several analysts note that the medium and long-term prospects are dire, and that either new resources will be pooled into defence or most of the ambitions of French defence policy will be unrealized.[41] Besides equipment, also the readiness of French forces is already well below the requirements established by the key doctrinal documents (Ministère de Finances, 2010).

In the case of Italy, the evolution of the budget calls its sustainability over time into question. At the beginning of 2012, the government undertook a spending review of the defence sector. The early 2012 defence reform (*New Defence Model*), adopted with bipartisan consensus and enacted amid a very severe overall budget crisis, was a quite pragmatic attempt to remedy these problems by reducing personnel (from 190,000 to 150,000 military units in the following years) and large procurement commitments such as the F-35 in order to refit the armed forces to a new context of large budget cuts throughout the public sector in Italy. Simultaneously, economic constraints have led to a reduction of the number of troops involved in international operations, which is at the lowest level since the mid-1990s. This drop is also due to international transformations linked to the greater difficulties of regional and multilateral organizations in obtaining troop contributions to peace support operations. The reduction of Italy's presence abroad (especially since its withdrawal from Afghanistan) will relieve forces from the overstretch of the past years, but it will also directly diminish the benefits associated with a broad international presence.

41 Interview: de Durand.

The scale of operations that can be performed in relative autonomy is of course the main difference between the UK and France, on the one hand, and Italy on the other. Although defence budgets have some similarities in terms of patterns, they remain structurally diverse in both size and composition. The stratification over time of these differences has made the Italian armed forces unable to close structural gaps.

Two other stark differences emerge. The first that comes to attention when considering the simple quantity of documents produced in Italy, France and the UK is the limited number of papers produced by the Italian Ministry of Defence. In fact, the overall amount of information and strategic assessments provided by Rome is extremely small in comparison with the output by the French and British ministries. For instance, contrary to the Italian case, several documents also dealing with operational and tactical aspects of doctrine of the various military services are usually provided. Secondly, Italian documents seldom refer to knowledge and understanding gained directly from experience on the battlefield. While British and French papers openly report concrete lessons learned from the most recent military interventions, the various editions of the Italian *Libro Bianco* almost entirely omit explicit references to battles or to tactical and strategic approaches adopted during national missions abroad. Instead, they deal generically with the 'Italian way to peace support operations'. In this regard, the legacy of constant attention to the military as an instrument of policy in France and UK inevitably plays a key role in explaining this difference. The fact that Italy did not have a proper and autonomous defence policy in the Cold War era is the main distinguishing factor among countries. Despite the above-mentioned transformation of the post-bipolar Italian forces, overall public concern with the armed forces remains exceptionally limited in cultural, historical and political terms.

Chapter 4
Force Deployment and Field Experience from Afghanistan to Libya

This chapter analyses major military missions undertaken by Italy as the second set of empirical references with which to interpret defence transformation. The sections examine force deployment in the principal operations carried out since 2001: Afghanistan, Iraq, Lebanon and Libya. The focus is on the type of forces deployed and the approach adopted. Rather than providing a complete account of the missions, the sections empirically observe the coherence of force deployment with the type of mission, the process of adaptation to the environment based on existing doctrines and on learning in the field (which will be further developed in Chapter 5), and the crucial equipment deployed on the ground. For each mission, the sections focus on Italy and provide comparative references when appropriate. Each section contains references to French and British experiences in the same scenario. This chapter is based on a comprehensive analysis of secondary sources as well as on extensive primary ones, such as official documents and in-depth and semi-structured interviews with experts and military officers who participated in planning, missions and training.

Chapter 3 has already illustrated the key features of the transformation of forces according to Italian strategic doctrine. These factors (substantially shared by France and Great Britain) are:

- Mobility and power projection capability;
- Holistic/comprehensive approach;
- Jointness and multinational interoperability;
- Digitalization and force protection.

Some additional clarifications are required. First, Italian defence has strongly emphasized its renewed power projection capabilities after forty years marked by a Cold War static approach. After the fall of the Berlin Wall, Italy started to deploy soldiers overseas, finally becoming a security provider. Therefore the need to increase readiness and (air and maritime) mobility is a crucial aspect shared by all strategic documents. Second, a multidimensional, holistic and 'comprehensive' approach clearly emerges as the main guideline in contemporary operations, which have required a combination of civil and military capabilities to reach the end-state of the mission.

Thirdly, jointness and multinational interoperability represent the cornerstones of the Italian doctrine in the post-Cold War era, when Italy has been constantly engaged in multinational operations. As we have illustrated in Chapter 3, a stronger integration with allies (and also among national services) is widely perceived as vital for the evolution of Italian defence. Finally, the strategic papers drafted in the new century have constantly focused on technological advances, digitalization and, above all, net-centric capabilities as essential requirements for the Italian armed forces. On looking at the evolution of the national doctrine in the last decade, 'force protection' can be coupled with 'digitalization' due to the growing importance of equipment to safeguard the lives of the troops.

The previous chapter also highlighted how the UK and France have shared these factors in their doctrines, prioritizing jointness, force protection, power projection capabilities, net-centric warfare, and a comprehensive approach. After assessing the overall level of coherence between budget and doctrine, the next sections focus on the consistency between the above-mentioned key elements and the Italian military operations undertaken since 2001.

Afghanistan: the other side of COIN

Italy at War?

The operation in Afghanistan is the most important mission that Italy has undertaken since the end of WWII. According to the Annual Army Reports, in 2010–2012 Italy suffered 27 casualties in military interventions abroad, 26 of them occurring in Afghanistan.[1] The human and economic costs of the mission were massive, and so were the military efforts undertaken on the ground, where Italian troops were involved in the riskiest combat operations in decades. In addition, at a broader level, the International Security Assistance Force (ISAF) was the first mission out-of-area carried out by NATO in its history.[2]

Italy contributed to both ISAF and Operation Enduring Freedom – OEF (which began on 7 October 2001 and ended in December 2006). While the Italian Parliament formally authorized the Italian military involvement at the beginning of November 2001, the small Italian contingent (around 380 soldiers) was deployed on Afghan soil only in February 2002, within the Kabul Multinational Brigade.

Italy made its most significant contribution to OEF through the Nibbio combat operation (February-September 2003). The 1,000-unit contingent was based in the Patkia province, patrolling the border with Pakistan in collaboration with the US and Afghan troops and engaging the enemy in a very complex scenario. Some authors have critically described Nibbio as a 'light force', given the absence of helicopters

1 See *Army Report* (2010), *Army Report* (2011) and *Army Report* (2012). In 2010–2012 the Italian soldiers wounded in military operations abroad amounted to 147 – 138 of them occurred in Afghanistan.

2 In 2003 NATO began to guide the operation. See the UN Security Council Resolution no. 1510.

and armoured vehicles (Gaiani, 2007; Ignazi, Giacomello and Coticchia, 2012). 'Without helicopters, heavy weaponry, or armoured land transport [...] the bulk of the efforts consisted of setting up checkpoints, establishing blocking positions at potential insurgent escape routes, and conducting intelligence-gathering patrols' (Schmitt, 2012, p. 7).

The mission was undertaken by light infantry battle groups.[3] Task Force Nibbio 1 was composed of Alpine troopers and Paratroopers, *Carabinieri*, NBC groups, a Heavy Mortar Company, corps of engineers with EOD (explosive ordinance disposal) capabilities, antitank platoons and Special Forces (Battisti, 2012). The Task Force was structured around two forward-operating bases (FOBs): Salerno and Chapman (Battisti, 2007). The mission, which ended in 2003 without casualties, was very brief. By contrast, ISAF was prolonged in its mandate and gradually extended its control over the country, which was divided into several Regional Commanders (RCs). In 2005 Italy became the leading nation of the Western area, composed of four provinces: Herat, Farah, Badghis and Ghor (Bertolini, 2009).

The aims of the mission were implementation of the Bonn Agreement, promotion of security, political stability, and institutional and economic development. The strengthening of security and governance and economic reconstruction were the main purposes of the intervention (Chiapperini, 2012). Although Italian forces were not deployed in more dangerous southern Afghanistan, the troops faced mounting security challenges in the weak and strategic western region, especially after 2005–2006. As testified by the numbers of casualties and wounded, the level of military confrontation progressively intensified: while between 2002 and mid-2005 three soldiers died, in the following years the number of victims increased considerably, reaching the figure of 47 at the end of 2012.[4]

In response to these growing threats, the Italian contingent was gradually enlarged. Whilst the troops numbered around 1,200 in 2006, the Italian cabinets (under pressure from NATO) sent more soldiers and equipment in the following years: in 2012, the vast majority of military personnel abroad were deployed in Afghanistan (around 4000 units). Additional armoured vehicles and aircrafts (such as AMXs and Tornados) were provided. In other words, Afghanistan clearly became a priority for Italian defence.

The Italian armed forces led the multinational contingent in the Regional Command West, where they engaged in the most intensive battles since the end of WWII. Already in 2007 the Operational Mentoring and Liaison Teams (OMLTs) were involved in fiercely-fought battles, and the Italian Special Forces (with the support of national intelligence and allies within NATO Task Force 45) operated at the border with Iran. But it was only after 2008–2009 that combat activities significantly increased, doing so in parallel with the US surge promoted by the Obama administration. With reduced caveat and limitations, Italian troops engaged in several battles (especially around the FOBs in Farah and Bala Murghab), also

3 See *Global Security* (2003).

4 For a detailed account: http://icasualties.org/oef/Nationality.aspx?hndQry=Italy.

suffering casualties due to suicide attacks. Responsibility was gradually handed over to the Afghan National Security Forces (ANSF), which gained increasing territorial control.

Overall, because of the high level of violence, the military intervention in Afghanistan 'has been the most demanding operating theatre, given the volume of personnel and activities it entailed' (*Army Report* 2012, p. 19). The Italian contingent, which will be withdrawn at the end of 2014, has also operated outside the RC West, mainly in Kabul (through a Battlegroup within the RC Centre under the French Command) and at the airport of Mazar-e-Sharif (with the Devil Task Force and Tornado aircrafts).

According to the doctrine examined in the previous chapter, armed forces should be expeditionary, joint and net-centric; and they should adopt a comprehensive approach to crisis operations. How and to what extent do the above-mentioned key elements consistently emerge in the evolution of the Italian military operation in Afghanistan? The next sections specifically focus (in comparative terms) on those crucial aspects concerning the cases of ISAF and OEF.

Mobility and power projection capability

Despite problems, mainly due to logistical setbacks,[5] with the deployment of the Air Force at the very beginning of the operation, after 2002 Italian armed forces proved to have considerable power projection capabilities. Already after parliamentary approval of the mission, the Navy was engaged within OEF with significant assets such as frigates and the aircraft carrier *Garibaldi*.[6] In terms of mobility, in 2005 the Air Force deployed the Support Mobile Unit (RMS) with the task of 'ensuring operational and logistical support to the flight units and mobile units operating outside the area'.[7] RMS also operated within the framework of the Aquila Task Force, establishing the first Forward Support Base (FSB) in Herat.

In the Regional Command West, both expeditionary logistic units (National Support Group – NSG) and enablers (such as the Army Aviation unit equipped with CH-47, A-129 and NH-90 helicopters), supported the troops and their readiness and mobility. Notwithstanding the important role played by the Navy

---

5 The Air Force played a limited role in the first months of the operation was also constrained by strict caveats on air strikes. For a detailed analysis see Gaiani (2007). The deployment of the contingent was strongly supported by the United States through several missions by C-17. Moreover, also the air supply of the FOB during the Nibbio operation was guaranteed by the USAF. However, multiple missions by Italian C-130J contributed to the mobilization (Battisti, 2012).

6 In addition, NATO's 'Active Endeavour' operation (which was undertaken in the Mediterranean Sea) was part of Operation Enduring Freedom.

7 See the official report on the operation in Afghanistan released on the website of the Air Force: http://www.aeronautica.difesa.it/Operazioni/Internazionali/afghanistan/Pagine/home.aspx.

and Air Force, the Army made the main contribution to the national operation (around 80% according to the Army Report 2012).

During Operation Nibbio, as reported by the then Commander Bertolini (2012), Italy adopted the 'Ground Assault Convoy' technique, a clear example of the growing need to increase effective and rapid deployability for transport over long distances (as in the Patkia province). In addition, in May 2003 the Nibbio Task Force undertook (for the first time in the history of the Italian Army) an air assault at battalion level during the Unified Venture operation[8] (Bertolini, 2012). Constantly stressed in the Italian strategic documents, the mobility of forces was confirmed as a crucial requirement for troops by OEF and ISAF, which developed new techniques and approaches in order to increase tactical mobility and the degree of readiness in the field.

Comprehensive approach

ISAF was the sternest test for the much-debated 'comprehensive approach' because of the challenges raised by security threats and compelling political and economic demands. Has Italy really adopted a 'holistic approach' on the ground? The answer is affirmative, also on considering Operation Enduring Freedom. Nibbio occurred years before the widespread diffusion of (NATO and US) doctrinal attention to non-military aspects of contemporary international missions. These elements already represented important guidelines, but they became leading principles for ISAF only in following months and years, mainly through the influence of the so-called Petraeus Manual (US Army and Marine Corps' *Field Manual 3–24*) drafted in 2006. At the border with Pakistan, Italian troops focused primarily on establishing relationships with local population and with the elders in the main villages,[9] displaying differences with respect to the more rigid US search and destroy approach. '[Italian soldiers] participated in Operations Haven Denial and Warrior Sweep with the 82nd Airborne Division, Afghan National Army and Special Operation Forces. They conducted more than 2,000 conventional patrols that led to the disposal of many weapons caches, and they developed a strong relationship with local Afghans'.[10]

In general, the attention devoted to reconstruction and development is a widely recognized feature of Italian military operations abroad in the post-Cold War era (Ignazi, Giacomello and Coticchia 2012). Detailed consideration of the overall Italian engagement in Afghanistan identifies two salient features of the

8 By contrast, the above-mentioned nature of Nibbio as a 'light force' had several consequences. The USA had to provide support through attack helicopters (which were lacking) and air supply for Italy's bases. Only at the end of the operation did Italy employ C-130Js. See Coticchia and De Simone (2014).

9 Interviews: Officer A, Folgore Parachute Brigade, Officer B, Folgore Parachute Brigade.

10 See Global Security (2003).

evolving national 'reconstruction-oriented' approach that fit well with the 'holistic perspective' constantly mentioned by the doctrine.

First, also due to the nature of its troops (the *Carabinieri*, a mixed police and military force and the Army) and the expertise accumulated in previous decades, Italy played a significant role in the training of Afghan security forces[11], both at the Regional Training Centre in Herat and at the Central Training Centre in Kabul (and also within the framework of the European Union Police Mission).[12]

Contrary to what had happened in the past, the *Carabinieri* adapted their traditional training to the combat scenario of Afghanistan, paying closer attention to the military dimension. The process of 'Afghanization' illustrates the centrality of training for the entire ISAF mission, which considered the quantity and the quality of the trained local security forces as one of the key benchmarks of the operation's success. In this regard, a typical asset of the Italian armed forces (i.e. the training of local forces) became a vital feature of the 'comprehensive approach' developed in the post-2001 military operations abroad.

Second, the combination of civil and military units, the so-called CIMIC, which Italy had already developed at the beginning of 1990s in Somalia with the COU – Circuiti Operativi Umanitari (Coticchia, 2013) – enhanced its importance in Afghanistan. As pointed out by the Army Report: 'military efforts could be combined with civilian, economic, political support, and with reconstruction as part of a wider security framework' (2012, p. 19). Through the units of the CIMIC Group South (CGS) Italy supported local authorities in the construction or rebuilding of public utility facilities with the general aim of promoting development and restoring security conditions. In the case of Afghanistan, Italian CIMIC forces became an essential component of the Provincial Reconstruction Teams (PRTs) in the Regional Command West (RC-W). They sought to enhance local political and economic conditions through several development projects and programmes. Italy led the first PRT in Herat in 2005. According to the Italian documents, PRTs are 'small teams composed of military and civilian personnel that operate across Afghan provinces. They provide security for relief operations and support humanitarian assistance or reconstruction in areas where the conflict is still ongoing or is marked by a significant degree of insecurity'.[13] Also an official

11 The training of Afghan security forces has been generally provided by the OMLT (later Police Advisor and Military Advisor) Teams of the Italian Army since the very beginning of the operation. In 2008 Italy assumed the leading role of three important training centres (in Adraskan, Herat and Kabul) for Afghan Uniformed Police, Afghan National Police and Afghan National Civil Order Police. Interview: Staff officer A, General Staff of the Army.

12 Both Carabinieri and Guardia di Finanza (Custom and Tax Police) have trained Afghan local security forces (national civil order police and border police) within the EUPOL framework. See: http://www.eupol-afg.eu. Italy supported the operational development of the Afghan National Security Forces (ANSF) also through the so-called Advisor Teams (AT).

13 *Army Report* (2012, p. 20).

diplomatic representative of the civilian component was attached to the PRT, which coordinated all the activities undertaken by national and international actors.

Such a combination of military units and civilian groups of experts was not a novelty for Italian operations abroad, but the PRT (as well as training) became a crucial component of NATO's comprehensive approach and of the COIN gradually developed by ISAF. According to the former Chief of Staff of ISAF, General Bertolini (2009), the multinational contingent openly adopted the doctrine developed by Petraeus in 2006 (and then applied in Iraq). For instance, new metrics were adopted to assess the progress of the mission in conformity with new COIN indicators. An effort was also made to achieve new benchmarks concerning the training of local security forces (Cordesman, 2012). While Bertolini emphasized the innovation already introduced by then General Commander of ISAF McKiernan in his Tactical Directive of December 2008, the Italian troops perceived the real turning-point of the mission in the new approach devised by McChrystal.[14]

Italy developed its traditional assets within this framework. The PRTs were conceived as primary tools in a counterinsurgency strategy because of their capacities to promote economic and social development. The Italian perspective on the relationship between COIN and the comprehensive approach' has been well clarified by the former commander of Regional Command West, Luigi Chiapperini, who views ISAF as a 'Counterinsurgency Full Spectrum Operation', a mix of civil and military efforts 'where a "comprehensive approach" should be developed' (2012, p. 103). Accordingly, both PRTs and training are vital in order to support local authorities. As remarked by Bertolini, Italy has made a major contribution to the COIN strategy developed by NATO, which has had four main tactical objectives: 'shape, clear, hold and build' (2009, p. 48).

For instance, the 'ink spot strategy' aimed at creating broader areas of security was increasingly adopted by the Italian task force through a growing number of FOBs in the Western Area.[15] Italian force provided also 'Combat Out Post' (COP) especially in the rural areas.[16] A more 'dynamic approach' to the bases, which were constantly upgraded and enlarged in order to maintain stable control of the territory, also emerged in the Western Area (Battisti, 2009).

Digitalization and force protection

The link between two of the main factors emphasized by Italian doctrine (digitalization and force protection) is well illustrated by the evolution of the Italian mission in Afghanistan.

As already stated, at the very beginning of ISAF and OEF, Italy's contribution was limited, with the exception of Nibbio. Especially on comparing the first years

14 Interviews: Officer A, Folgore Parachute Brigade, Officer B, Folgore Parachute Brigade.

15 Interviews: Officer A, Folgore Parachute Brigade, Officer B, Folgore Parachute Brigade.

16 Interviews: Staff officer A, General Staff of the Army

of the Italian military engagement with the subsequent involvement, the perception of the military is that of the ever greater deployment of equipment mainly due to the growing tasks and the mounting violence.[17] After 2006, the new weapons provided to the Italian Task Force had increasing net-centric capabilities intended to enhance the acquisition of information on the ground, such as unmanned air vehicles (like the Predator), ISTAR (Intelligence, Surveillance, Target Acquisition and Reconnaissance) teams or the VBLM Freccia vehicles with which a manoeuvre company was equipped for the first time (Army Report, 2010).

The Pinerolo Mechanized Brigade was the first NEC unit deployed on the ground to 'test' the net-centric capabilities by adopting the VBLM Freccia (armoured tactical multi-role vehicles) and the SICCONA C2[18] systems (Nones and Marrone, 2011). This engagement marked considerable progress in the shift of the armed forces towards greater digitalization.[19] From the perspective of the General Commander of the Brigade, the added values provided by the NEC capabilities were the accuracy and the promptness of information.[20] However, the operative involvement of the Freccia was not particularly intense because of the nature of the Afghan scenario, which was not considered a 'high intensity conflict'.[21]

As regards digitalization and force protection, specific attention should be paid to the counter-IED efforts made by Italy to improve the safety of its troops. The need for adequate protection became pressing after 2006 and urgent after 2008.[22] Hence armoured vehicles, UAVs, NH-90 and A129 Mangusta helicopters reinforced the Italian contingent (Coticchia, 2013). In particular, the Army's VTLM Lince (light tactical multi-role vehicles) provided the soldiers with a greater level of protection (especially compared with the previous VM90s[23]). After some dramatic incidents when machine gunners atop the vehicles died after attacks, the Italian forces were provided with a limited number of reinforced Lince with an armoured turret.[24] Moreover, instead of waiting for the modified versions of national equipment, in 2008 Italy adopted also US vehicles[25] in order to counter the growing menace

17 Interviews: Officer A, Folgore Parachute Brigade, Officer B, Folgore Parachute Brigade, Staff officer, NATO, experience as Commander of the Italian Task Force in RC-W, General Commander, Taurinense Alpine Brigade, Officer, NATO Rapid Deployable Corps Italy.

18 SICCONA is the *Sistema di Comando, Controllo e Navigazione* (Command, Control and Navigation System).

19 Interviews: Staff officer A, General Staff of the Army, Staff officer B, General Staff of the Army.

20 Interview: General Commander, Pinerolo Mechanized Brigade.

21 Interview: General Commander, Pinerolo Mechanized Brigade.

22 According to the interviewees who fought in Western Afghanistan, the escalation of violence and attacks against Italian troops occurred in mid-2011.

23 Interview: Officer, NATO Rapid Deployable Corps Italy.

24 See Ansa (2009). However, the heavy turret was considered risky if the vehicle overturned.

25 Interview: General Commander, Pinerolo Mechanized Brigade.

posed by the IEDs. Thus the Mine Resistant Ambush Protected (MRAP) Cougar and the Mine Protected Clearance Vehicle (MPCV) Buffalo MK2 were introduced on the ground.

Moreover, in the 'IED battle' Italy created Weapons Intelligence Teams in order rapidly to collect lessons learned from attacks, providing (after 48 hours) shared reports on the event. The Italian forces constantly adapted their approaches, and especially their equipment, to improve mine clearance. For instance, as revealed by a former Commander of a Weapons Intelligence Team, new metal detectors were introduced, or smaller (and more flexible) anti-landmine robots were employed to increase the overall effectiveness of the military efforts to counter the new tactics adopted by the Taliban.[26] The changes were made rapidly, as in the decision to install counter-IED jamming systems in almost all the vehicles in a convoy, rather than in only the first and last ones.[27] The transformation was not limited to equipment but influenced the Army's entire approach. For example, the corps of engineers evolved from providing a static support for manoeuvres to performing a more dynamic role required by the constant counter-IED operations.[28]

Jointness and multinational interoperability

Italian documents encompass the multidimensional perspective of contemporary security, and they clearly recognize the role played by multilateral institutions in fostering that view. As noticed, the influence of NATO and the USA on doctrine (i.e. the comprehensive approach and COIN) has been massive: in particular Italy has derived most of its recent doctrinal reflection directly from the Atlantic Alliance.

In terms of deployment, Afghanistan has probably been the crucial event for enhancing multinational interoperability at intervention level. Constant cooperation among allied forces has been a distinctive feature of OEF and, especially, ISAF. Procedures, approaches, equipment and lessons learned have been regularly shared among partners, especially since 2005–2006 with the worsening of security conditions.

In regard to Italy's military involvement in Afghanistan, we would emphasize four main aspects regarding jointness and multinational interoperability. First, at tactical level, Italian troops were continuously involved in military actions together with allies and Afghan security forces, fighting the most intense battles of the post-WWII era. Joint combat operations were undertaken from the outset of the mission, as had already occurred in 2003 in the cases of Dragon Fury or the previously cited Unified Venture (Battisti, 2012). But, as recognized by the General Commander of the Folgore Parachute Brigade, it was mainly during ISAF

26 Interview: former Commander of a 'Weapon Intelligence Team', Taurinense Alpine Brigade.

27 Interview: Officer, NATO Rapid Deployable Corps Italy.

28 Interview: Officer, NATO Rapid Deployable Corps Italy.

that joint operations gave an 'enormous impetus' to interoperability with the allies, especially as regards command and control and targeting.[29]

The deployment is here conceived as the main driver of growing multinational interoperability, specifically at the tactical level, with a significant enhancement of the standardization of sharing procedures.[30] It is also possible to track remarkable evolution of the degree of multinational interoperability during the years of the mission: whilst in the case of Nibbio the Italian Special Forces (SFo) were not assigned to the Combined Joint Special Operations Task Force.[31] After 2006 the Italian SFo (mainly *Col Moschin* paratroopers) played a very active role in the joint Task Force 45, under the direct command of the COMISAF. As stated by a previous Italian Commander of Task Force South, actions in the troubled area of Farah were constantly planned with the allies.[32] The effort was remarkable, especially for the Army (while the Air Force had been more integrated during the Cold War, as illustrated by the anti-aircraft artillery that already required such interoperability).

Second, interoperability was not limited to relationships with the US or the main European armies. In some cases, multinational interoperability required involvement in operations alongside 'minor allies'. For instance, Italians complained about the lack of combat support (helicopters, corps engineers) from the Lithuanian contingent.[33] Moreover, specific attention should be paid to the training and mentoring of Afghan security forces through the OMLTs (Operational Mentoring Liaison Teams). In this case, the joint efforts encompassed shared involvement in combat operations. For instance, as reported by the Former Chief of Staff of ISAF, Italy was the third largest contributor to OMLTs in 2009 (Bertolini, 2009).

Third, above the tactical level, ISAF provided (largely thanks to McChrystal) a structured intermediate joint command dedicated to the operational dimension. The multilateral framework offered standardized procedures and guidelines (such as those on cultural awareness or mitigation), promoting (mainly from the top-down) a constant integration among the allies (especially after 2009).[34] Moreover, a crucial function was performed by the Afghanistan Mission Network, which 'creates a common operating picture for all U.S. and NATO commanders in

29 Interview: General Commander, Folgore Parachute Brigade. According to the General, the more tactical capabilities are developed, the more multinational interoperability is effective.

30 Interview: General Commander, Folgore Parachute Brigade.

31 On caveats, RoEs and, above all, political restrictions during Operation Enduring Freedom (and ISAF) see, among others: Gaiani (2007) and Ignazi, Giacomello and Coticchia (2012).

32 Interviews: Officer, NATO Rapid Deployable Corps Italy, Staff officer, NATO, experience as Commander of the Italian Task Force in RC-W.

33 Interview: Staff Officer, General Staff of the Army.

34 Interview: Officer B, Folgore Parachute Brigade.

Afghanistan. [The Allies] have subsumed their own internal secret communications networks in favour of a common network for managing command and control and sharing intelligence, surveillance and reconnaissance information'[35].

Fourth, Italy occupied important positions in the international ISAF chain of command. Thus also above tactical and operational level, ISAF promoted a broad idea of interoperability at the highest level. On several occasions, Italian generals assumed crucial responsibilities within the multilateral framework. For instance, the Italian general Mauro Del Vecchio was in charge of ISAF between 2005 and 2006. Such efforts enabled the achievement of new procedures, competences and lessons learned regarding the joint command (Del Vecchio, 2012).

UK and France in Comparative Perspective

The above-mentioned key elements have affected the military operations undertaken by French and British forces in Afghanistan. What other aspects emerge on considering their long-standing involvement in the war in light of the main features of military transformation illustrated in the previous chapter?

Starting from the end, we emphasize that the conclusion of ISAF will certainly represent a turning point for the British defence, but especially for the Army. Indeed, the operation was an 'insurance' for the Army, helping it to limit the impact of financial cuts due to the massive involvement of troops on the ground.[36] Moreover, the approach adopted in Afghanistan (mainly by the Army) in order to counter guerrilla warfare seems more consistent with a political context marked by austerity and financial constraints.[37] In fact, apart from new technologies such as UAVs for acquiring surveillance capabilities or counter-IEDs tools, COIN is 'people-intensive' rather than 'equipment-intensive'.[38] However, after the withdrawal of troops the British army will certainly suffer cuts in personnel that can probably be offset by investments in reserve units.[39]

Four main aspects emerge on looking at the British military involvement on the ground and at the process of adaptation and transformation of forces.[40] The first and most general aspect is the level of commitment of the British forces deployed in 2006 to Helmand province in southern Afghanistan, one of the most dangerous

35 See Rosenberg (2010).

36 Interview: Chalmers.

37 Interview: Taylor.

38 Interview: Taylor.

39 Interview: Farrell.

40 A further feature to be noted is the large amount of research, data, and empirical analysis on the UK mission in Afghanistan. In this regard, the British public and academic debate is not at all comparable with the Italian one, which has paid very limited attention to the national military commitment in Central Asia.

areas in the entire country.[41] Indeed, the fighting in that region was the most intense that British forces had experienced since the Korean War (King, 2010). The UK paid a high price in terms of casualties, losing 453 soldiers in Afghanistan, most of them after 2008.[42] London notably increased its forces after the first deployment in Helmand, doing so from 3,500 troops in 2006 to 19,000 in 2010.

The second, and probably the most notable, factor is the evolution of the British counterinsurgency strategy. As stated by Egnell, British doctrine and historical accounts have provided lists of several key principles, such as political primacy, close civil–military cooperation, minimum force, and flexibility' (2011, p. 300). Oddly enough (as we will discuss later in regard to the Iraq War), the UK was not able to organize an effective counterinsurgency campaign for several years in post-2001 conflicts (Farrell and Gordon, 2009). The British belief of having much to teach rather than to learn collapsed on contact with the enemy in Afghanistan (Little, 2009). In 2006 the British contingent in Helmand was essentially a small peacekeeping force (King, 2010). Then, after a dramatic worsening of security conditions, the British campaign in Helmand evolved 'from one centred on hard military power and directed at destroying the Taliban, to one focused on generating "soft effects" and securing the civilian population' (Farrell and Gordon 2009, p. 673). According to King (2011), after the Iraqi failure the British troops conceived Helmand as an opportunity to prove their values. However, with an insufficient force[43] and poor intelligence, they adopted excessive force, heavily relying on firepower, aligning with corrupt local leaders and thus alienating the population (Farrell and Giustozzi, 2013). Only after the end of 2007 did the situation begin to change, also because of the new approach fostered by McChrystal, who redefined the COIN principles (chiefly protecting population and enhancing the capabilities and the Afghan government and local security forces), focusing more effectively on stabilization and development, rather than concentrating on defeat of the enemy (Farrell and Chauduri, 2011).

Also, a population-centric strategy was implemented with some positive progress (although limited to the operational level alone). As stressed for Italy, also for the UK the 'comprehensive approach' played a major role in fostering better understanding of the local context as well as the importance of the non-military dimensions. Likewise, the British concept of effect-based operations

41 'Operation Herrick' began in 2002. In April 2006 it was followed by 'Herrick 4', escalating the British commitment to ISAF.

42 See: http://icasualties.org/oef/

43 As well illustrated in the literature (Farrell and Gordon, 2009; King, 2010) one of the main errors committed by the British forces at the beginning of the deployment in Helmand was the dispersion of forces, which were isolated among scattered FOBs with few troops. Dramatic battles, such as those of Musa Qala, Sangin and Now Zad exposed the flaws of this approach (which was contrary to the traditional strategic principle of concentration).

(EBO), which was at the centre of the strategic documents, had the same function of improving understanding of the local context (Farrell 2008).

The third feature concerns the British capacity to provide the equipment required on the ground. The troops initially suffered from shortages of military apparatus (primarily attack helicopters and vehicles).[44] According to King (2010), at the beginning of the Helmand campaign the British troops far exceeded the resources available. However, the UK was able to adapt to the security challenges posed by the operation: more troops and resources were provided. Indeed, the contingent was better equipped on receiving weapons (such as new body armour, rifle sights, Mastiff armoured vehicles, GMLRS – Guided Multiple Launch Rocket System) largely acquired as Urgent Operational Requirements (King 2011; Farrell, Rynning and Terriff, 2013). As in the case of Italy, the protection of personnel became a priority in the procurement process.[45]

Fourth, also for the UK the military operation in Afghanistan marked significant progress in terms of jointness and multinational interoperability. For instance, Farrell and Chauduri (2011) emphasize the importance of intermediate operational headquarters in the ISAF joint command.

As said, the influence of the multilateral context was considerable, especially at the doctrinal level, where the UK 'developed the brain it had lost until 2007',[46] principally due to the impact of the Petraeus and McChrystal approaches.

However, three main obstacles to wider interoperability persisted: scant coordination among regional commands, the ethos of regimental culture (for elite paratroopers) based on the ideal of fighting independently of other forces and, above all, inter-service rivalry (King, 2010). As well illustrated by Cornish and Dorman (2009) regiments want to demonstrate their skill because operational performance is crucial for institutional survival in a context of limited resources.

The case of France evinces a quite different pattern in the national mission in Afghanistan. The shift in France's commitment occurred in 2008, when it assumed the Regional Command Capital in Kabul and, above all, extended its responsibility to Surobi District, east of the capital, and Kapisa province. Also France shaped its forces according to the above-mentioned features of FOBs and OMLTs (which were also deployed in the southern Uruzgan region).

---

44 See, for instance, Towned (2009), and Major-General Binns, oral evidence, *Iraqi Inquiry,* 15 January 2010. According to Farrell, Rynning and Terriff (2013) the MoD then adequately provided vehicles, while shortages of helicopters remained.

45 As revealed by King: '[the] Mastiff armored vehicle, was originally acquired for Iraq within five months under the UOR scheme; some of these units were redirected to Afghanistan. Delivery of a larger purchase of Mastiff for Afghanistan is taking longer because of a complex refitting of the vehicle to meet campaign-specific specifications combined with limitations on industry's ability to meet demand In addition, in October the Defence Secretary announced a £500 Million "Protected Mobility Package" providing an additional 500 armored patrol and support vehicles for operations in Afghanistan' (2011, p. 677).

46 Interview: Farrell.

The Kundau ambush of August 2008 (when 10 soldiers died) altered the status quo in the area under the French command.[47] The French casualties, mostly from the 8th parachute regiment, represented the largest loss of life suffered by the French army since a bombing in Lebanon in 1983, and it was also the deadliest attack on international troops in Afghanistan in more than three years.[48] In the entire French operation, Paris suffered 86 casualties, and only 14 of them occurred before 2008.[49]

In April 2009, after it was decided to reinforce the French contingent, 700 additional units were deployed to Afghanistan. Indeed, the pattern of the French presence in Afghanistan was affected by various changes. The national commitment under OEF and ISAF was limited (being mainly based on training activities in Kabul). Despite the substantial contribution made by the Air Force and Navy (Shapiro, 2002), the 1,200 French troops deployed after 2002 were not involved in open combat against the Taliban; the main focus was on training and stabilization.[50] According to Saideman and Auerswald (2012): 'The deployments of French forces in Afghanistan have not followed patterns consistent with the rest of NATO. When NATO rolled out of Kabul and spread across Afghanistan in 2005–2006, the French conventional contingent remained in Kabul […]. The restriction to Kabul essentially prevented the French battalion from engaging in combat' (2012, p. 78).

However, France then modified its approach, first sending OMLTs to Uruzgan[51] and later deploying its forces to the (dangerous) eastern province of Kapisa, thus allowing the redeployment of some American units from East to South. President Sarkozy wanted to provide greater commitment to the ISAF, also as part of broader involvement in NATO, which reintegrated the French forces into the command structure after decades. However, at the end of 2012 the new president Hollande announced the withdrawal of French forces before the end of the ISAF.

---

47 Heated debate arose in the aftermath of the dramatic ambush, indirect responsibilities for which were attributed by the media also to the Italians. Indeed, according to some reports (always denied by Rome), the Italian forces paid bribes to the Taliban in the same area in order to avoid attacks. The end of this alleged tactic was considered to be one of the primary causes of the ambush. See, among others Foust (2012).

48 For more details see Burke (2008) and Keaten (2008).

49 See http://icasualties.org/oef/Nationality.aspx?hndQry=UK.

50 As reported by Shapiro (2002), almost a quarter of the French Navy contributed to the very beginning of the operation, providing a naval task force led by the aircraft carrier Charles De Gaulle. During Operation Anaconda in March 2002, French Mirages became the only non-US jets to conduct strike operations in Afghanistan. On the French Air Force in OEF see Lewis (2002). Also French Special Forces were involved on the ground but their withdrawal was announced in 2007, before Sarkozy adopted a different commitment.

51 Like Italy, also France had strong caveats and strict rules of engagement. The allies complained about the delays due to the need for permission from France. See, among others: Saideman and Auerswald (2012).

In summary, consideration of the various phases of the French mission highlights four main aspects in terms of military transformation. First, significant power projection capabilities are evidenced by the contribution that the French Navy and Air Force made to the international mission from the outset (Shapiro 2002). Specifically in regard to the mobility of forces, the Caracal and Gazelle helicopters, which gave ISR support to French troops, played a very important role. As stated by the Colonel Perrin (2009), a former commander of a French battalion, the airmobile tactics enabled the crew to maintain constant initiative and surprise. Moreover, 'the fire support provided by the helicopter is much more precise than ammunition delivered by planes. In a country where any civilian loss pushes the rest of the family into the Taliban's arms, this point is particularly important' (2009, p. 53).

This last statement is closely related to the second crucial aspect: the French COIN. As previously illustrated, the political perspective adopted by Chirac towards the mission limited effective French contribution to operations on the ground. This view directly affected the French strategy, which showed an initial 'aversion to PRTs' (Saideman and Auerswald, 2012, p. 78) since they implied a longer commitment in the theatre, putting troops in danger. Thus kinetic activities were mainly restricted to the (almost invisible) action carried out by the Special Forces. In this first phase, France also emphasized its disagreements with the strategy of the international operation, especially with the excessive use of military instruments to the detriment of diplomatic and economic tools (Shapiro, 2002). This stance changed under Sarkozy, who fostered a more significant military commitment implementing already-mentioned principles such as the so-called 'ink spot strategy' aimed at increasing the level of security and control of an area. However, at the beginning of this second phase, France made some of the mistakes that it had previously denounced as American flaws: principally an aggressive approach focused on fighting the Taliban in Kapisa province (Foust, 2012).

Only after 2009 did the French forces soften their approach, focusing more closely on protection of the local population, and trying to improve political, social and economic conditions in accordance with the views expressed by the Forces Employment Doctrine Center (Coste, 2009). France enhanced the role of civil-military activities (which were gradually cut back in 2008) taking advantage of US CIMIC support in Kapisa province. Thus, the French armed forces repeatedly modified their approach on the ground, following the different political courses taken by governments. However, the transformation also related to a more dynamic interaction between Paris and the context on the ground, especially in regard to the third main aspect: IED counter-warfare and force protection. The Joint Doctrine Centre started to work on IEDs in 2005, without developing a specific structure (as the British did) but only creating a working group supervised by the Joint Staff. As stated by Lt. Colonel Girault (2012), France devised a constantly evolving doctrine on IEDs, exploiting the findings of the Weapon Intelligence Teams after incidents. A flexible approach integrating new views and capabilities was combined with a better-equipped force. Additional troops, helicopters, drones and detached electronic warfare assets sustained the military effort, also in IED counter-warfare. The crucial importance of engineer squads, as illustrated by the Italian case, is also evident in the French operation, which systematically structured joint teams with infantry platoons.

As for the CIMIC, the French contingent was supported by such allied means as air support and UAVs, especially for the counter-IED warfare (Perrin 2009). This introduces the last main aspect: the growing importance of jointness and multinational interoperability. Tactical engagement with allies and local forces (from OMLTS to special forces, also in southern Afghanistan)[52] as well as the command of multinational groups in Kabul reveal the increasing need to share information, capabilities and structures (such as Joint Improvised Explosives Device Defeat Organization) in order to achieve more effective outcomes on the ground in contemporary operations.

To sum up, besides the mentioned distinguishing features, jointness, multinational interoperability, force protection, mobility and 'comprehensive approach' were crucial aspects shared by three national contingents forced significantly to adapt and transform their approach to the conflict in Afghanistan. As illustrated in regard to the spread of a new COIN doctrine, strong influence on the above-mentioned changes was directly exerted by another front: Iraq.

Iraq: Tragedy and Retreat

Antica Babilonia

The Antica Babilonia mission (June 2003 – November 2006) was one of the most controversial and dramatic operations undertaken by Italian troops in decades. Italy suffered 33 casualties, 23 of them killed by bomb attacks. The Berlusconi government deployed a national 3000-unit contingent in the southern Dhi Qar province some months after the US military intervention began. As illustrated by recent studies, the operation was shaped and designed within the framework of a 'humanitarian emergency': according to the Italian decision-makers, the mission was essentially a peace operation aimed at providing humanitarian aid, repelling local bandits, and fostering development and reconstruction (Coticchia and Giacomello, 2007; Petrilli and Sinapi 2007). Despite the dramatic worsening of security conditions on the ground and the bloody attacks suffered by Italian troops (and contrary to the shift that occurred in Afghanistan during the second phase of the intervention), this 'humanitarian' approach remained constant until the new centre-left government withdrew Italy's forces in 2006. In a sense, Italy devoted specific attention to reconstruction, mainly through the CIMIC units, which played a role of great significance (especially due to the absence of NGOs

52 One of the most relevant examples is provided by the Chimere Task Force (with the USA) in Kapisa (Perrin 2009).

in the field). However, the hostile conditions in Nasiriya forced Italian troops to engage in several severe battles, most of them in 2004. According to Angeli (2005), a development expert who worked in Nasiriya for some time, Italian soldiers accomplished only a few sustainable reconstruction projects, despite their bold efforts, because of the downsizing of security.

Some analysts considered the original Italian equipment to be 'dramatically inadequate' (Gaiani, 2007, p. 56). Tanks, infantry-fighting vehicles or assault helicopters were not initially deployed. Several weapons were provided only after the violence against the Italian contingent openly erupted (Ignazi, Giacomello and Coticchia, 2012). The adjustment process was slower in Iraq than in Afghanistan.[53] Despite these problems, which were mainly related to the never-modified political interpretation of the mission, the national contingent was one of the largest in Italy's post-Cold War history, also encompassing the contributions of the Navy and the Air Force.[54] The Army and the *Carabinieri*, who created a specific Mentoring Monitoring Unit (CC-MMU), were deployed among several bases in Nasiriya, where they took advantage of manoeuvre units, logistics, RISTA and NBC defence. On looking at the Italian military engagement in depth, we can split Antica Babilonia into two crucial moments.

The first, and the most dramatic, one was the suicide attack of 12 November 2013 that killed 17 Italian soldiers and 2 Italian civilians. A truck-bomb destroyed the Italian Maestrale compound, whose protection was extremely weak (without an adequate system of passive defence, outposts with RPGs, defensive protections like Hesco Bastions) in preventing a possible VBIED (vehicle-borne improvised explosive device).[55] Apart from the technical inadequacies, the base's vulnerability was mainly due to its location in the city centre and its 'openness' to the civilian population as required by the interpretation of 'humanitarian intervention' (Ignazi, Giacomello and Coticchia, 2012).

The second key moment of the mission came in mid-2004, when the three intense so-called 'battles of the bridges' occurred (Cappelli, 2005). The operation Porta Pia involved three companies with the final goal of freeing the bridges of Nasiriya. The Shia's Mahdi Army militarily engaged the Italian troops, also assaulting the Italian bases and the Central Provisional Authority (Petrilli and Sinapi, 2007). Evident during these dramatic clashes was the initial absence of tanks (Ariete or Dardo), UAVs and mainly armoured helicopters (Mangustas, the lack of which hindered the more rapid transportation of wounded soldiers).[56]

53 Interviews: Officer A, Folgore Parachute Brigade, Officer B, Folgore Parachute Brigade.

54 As reported by Ignazi, Giacomello and Coticchia (2012), while the Naval Group was in the port of Umm Qasr, the Joint Air Task Group IRAQ (JATG-I) was in Tallil.

55 For a detailed analysis see Petrilli and Sinapi (2007) and Coticchia (2013).

56 Without armoured helicopters the rescue of a seriously injured soldier was extremely complicated. The death of this private of the Serenissima Regiment during the assault on the Libeccio base fuelled the polemic (Ignazi, Giacomello and Coticchia 2012).

The last months of the operation were marked by a growing deterioration of security conditions, as testified by IED attacks on Italian troops. When Italy withdrew[57] its contingent (in 2006, before the US surge), Iraq was devastated by a massive wave of sectarian violence. Despite the failure of the international operation until that year and the problems faced by the Italian contingent, several significant considerations can be made concerning the evolution and adaptation of forces (or lack thereof) on the ground.

Mobility and power projection capabilities

In terms of power projection capabilities, the most remarkable aspect of the Italian mission in Iraq is undoubtedly the well-planned and effective redeployment of the contingent: Operation Itaca. Notwithstanding massive concern, the withdrawal occurred without problems, evidencing the capacity of the Italian armed force to operate far from national borders with efficiency. In terms of mobility, despite shortages of equipment, the Italian Joint Task Force provided air support (also with the HH-35 helicopters) for reconnaissance missions, convoy escort and transport.[58]

Comprehensive approach

The mission Antica Babilonia accounts some important lessons learned in the field of Stabilization and Reconstruction (S&R). The worsening security situation that surrounded the intervention was a (dramatic) novelty for most of the forces involved. The mission was consequently a severe test for the Italian 'comprehensive approach'. Schematically, we can identify three main aspects.

First, the CIMIC structure (owing to the absence of most Italian NGOs, which were politically against the Italian involvement in Iraq) performed a prime role in almost all the aid projects. The CIMIC Group South, Task Force Livenza, implemented several programmes in diverse sectors, such as public health or culture. Anti-mine activity was once again a constant asset for Italian forces. Also in Iraq the PRT coordinated efforts devoted to development and reconstruction, seeking to gain the trust and support of the local population. However, the dramatic security scenario hindered the achievement of positive results.

Second, as in Afghanistan, the training of local forces (both the Iraqi Army and the Police Service) was a crucial part of the Italian mission, especially due to the specific nature of the *Carabinieri*. They were greatly appreciated by the

---

The Mangustas were also required for the protection that they provided. Several pilots openly complained about their absence (Coticchia, 2013). The controversy was exacerbated by the death of an Italian soldier after being shot while flying in an AB-412 non-armoured helicopter.

57 Also after the end of Antica Babilonia, around 80–100 *Carabinieri* remained in Iraq (in Baghdad) within the NATO Training Mission.

58 As illustrated by the official account of 'Antica Babilonia' provided by the Air Force: a helicopter unit operated at the airport of Tallil close to Nasiriya, the second largest and most important airport in Iraq.

allies: the USA asked Berlusconi's cabinet to manage the training of police forces throughout the country. However, the government refused in order not to appear excessively committed to the US intervention, which had been strongly opposed by public opinion (Gaiani, 2007; Coticchia, 2014). Overall, 'the most important lesson identified from Iraq-2003 was the urgent need to foster the reconstruction of local security forces as soon as possible'.[59] And Italy had the tools and assets with which to play a substantial role. However, the Dhi Qar province revealed how the ethnic and religious affiliations were obstacles to the creation of a 'national' police force. In Nasiriya the local security forces remained more loyal to Shiite parties than to the government, so that the 'Iraqization' process suffered notable setbacks.

Third, the 'comprehensive approach' was not systematized as it was in Afghanistan. The 'low profile' that characterized Nibbio was a constant approach that was never substantially modified, not even after the dramatic battles fought on the bridges of Nasiriya or around the national bases. In light of an effective 'national way to counterinsurgency' (before 2006 the term 'COIN' was almost never used by the Italian forces, especially because of the parallel framework of 'humanitarian mission'), the Italian intervention adopted a twofold approach. On the one hand, despite fierce clashes and violence, the Italians did not renounce dialogue and mediation with local militias (Petrilli and Sinapi, 2007). The allies openly criticized this soft approach, which was intended to prevent the degeneration of security conditions (as well as to avoid an open military confrontation that would have completely destroyed the 'peace mission' framework'). On the other hand, after the shocking suicide attack of November 2003 and the main battles fought in 2004, the Italian troops renounced control of the entire province by accumulating the contingent in bases outside the centre. The consequence of this decision, prompted by the above-mentioned causes, was that the militias gained greater control over the most problematic districts in the urban territory.

In summary, the lack of political will radically to alter the nature of the operation (notwithstanding the deployment of more adequate equipment) left Italian forces little space to adapt their approach to the challenges of the local scenario. Some traditional assets were provided (training and CIMIC) but the constraints (also related to the broader dramatic scenario of the conflict) remained considerable, while the decisions adopted after the main kinetic events did not prove particularly successful.

Digitalization and force protection

The initial force structure of Antica Babilonia had various flaws due to the lack of important equipment. Nonetheless, the mission testified to the increasing technological advances achieved by the Italian armed forces in recent decades. In a context of urban warfare, new digitalized capabilities (such as RISTA units or UAVs) guaranteed high flexibility and constant awareness and surveillance. As stated by the former General Commander Chiarini (2006), tools like drones

59 Interview: Officer, General Staff of the Army.

increased the safety of troops, who did not have to be involved in risky patrols on the ground to obtain information instead provided by the unmanned vehicle.

As recognized by the Air Force:

> The Predator is an unmanned reconnaissance aircraft, which, thanks to the performance of aerial surveillance and an extended flight range, is a valuable tool for the prevention and control of trafficking and potentially dangerous events [It] is an additional safety factor for the Italian and Iraqi military personnel engaged in joint missions for control of the territory.[60]

The Predator was one of the weapons sent *after* the beginning of the operation, being employed in the last two years of the operations for more than 1,600 hours.[61]

At the time of Antica Babilonia, drones were still a striking innovation. Worth noting is the surprise of the Italian military at the massive use made of drones by the Japanese for surveillance of their bases in Iraq.[62] This is probably a sector where 'the menace was faster than the attempts to adapt'[63]: it was evident that the protection of compounds was the most crucial aspect to revise and improve.

By contrast, counter-IED warfare evidences the capacity to modify approaches and tools, such as the 'rudimentary technical transformation' of the *Lince* to deal with threats, which were particularly severe in the last months of the Italian operation.[64]

Jointness and multinational interoperability

According to Schmidt, the deployment of troops in Iraq, even if problematic, considerably helped the evolution of Italian military instruments: 'Working with allies in a hostile environment far from Italy has forced Italian forces to 'up their game' when it comes to training, logistics, and field-level modernization' (2012, p. 7). On the one hand, Italy already had a trained and skilled structure to operate at joint level. For instance, logistics were already advanced in terms of interoperability, mainly as a consequence of the manifold multilateral operations undertaken in the Balkans.[65] On the other hand, Antica Babilonia was a further crucial step towards more effective multinational interoperability. For instance, training units were based on a joint component (with Romanian personnel). Also

60 Website of the Air Force on 'Antica Babilonia', available at: http://www.aeronautica.difesa.it/Operazioni/Internazionali/Pagine/Iraq.aspx

61 Website of the Air Force on 'Antica Babilonia', available at: http://www.aeronautica.difesa.it/Operazioni/Internazionali/Pagine/Iraq.aspx

62 Interview with Italian Senior Army Officer who was deployed in Iraq.

63 Interview with Italian Senior Army Officer who was deployed in Iraq.

64 Additional protection was added both inside and outside the vehicle to repel IEDs and EFPs (explosive former penetrators). Interview: Officer B, Folgore Parachute Brigade.

65 Interview: Officer, NATO Rapid Deployable Corps Italy.

the UAV Predator was a joint programme that required shared efforts by different services and manifold preliminary courses (that were held by the allies).[66]

Moreover, in regard to the chain of command within the multinational operation, the Chairman of the Joint Chiefs of Staff held operational command over national forces, while operational control was delegated to the Commander of the British Division stationed in Basra (Ignazi, Giacomello and Coticchia, 2012).

Overall, although Italy was able to provide the equipment (even the most technologically advanced) required to face the challenges raised by the evolving scenario, the initial design of the mission and the political intent of maintaining the inadequate frame of 'humanitarian emergency' considerably affected the promptness of adaptation and the final outcome of the intervention.

The UK in Iraq

A recent study defined the UK military operation in Iraq as a 'failure' (Farrell and Gordon 2009, p. 666). The fiasco was mainly blamed on the lack of understanding of political dynamics in Iraq and (contrary to what happened in Afghanistan) on an unwillingness to effectively learn and adapt (Mansoor, 2009). According to Iron: 'In 2005 and 2006, the British Army found itself countering a bitter insurgency in southern Iraq for which it was neither equipped nor trained to fight' (2013, p. 55). Similarly to the case of Italy, also the narrative adopted by UK leaders was misleading. In fact, at the very beginning, the British operation in Iraq (called Telic) was mainly described as a peace support operation, as 'nation-building', and rarely as COIN (Marston, 2009). 'The British army suffers from lack of soldiers, lack of money and lack of conviction'.[67] As reported by a harsh article in *The Economist*: the British commanders had 'lost their way'.[68]

In regard to the British engagement in Iraq, two main military operations carried out in Basra warrant specific attention: Sinbad (Autumn 2006) and Charge of the Knights (Spring 2008). From the rapid defeat (in only a few weeks) of the Iraqi armed forces in 2003 to 2006, when deadly sectarian violence erupted in Iraq (Ricks, 2009), British forces adopted a sort of 'peacekeeping approach' in the Basra province, which was under their control. Also Washington misunderstood the conflict, in the belief that it was dealing with terrorists rather than an insurgency (Kilcullen, 2009). However, after 2006 the allies took two different paths: the US adopted a new strategy based on surge and a proper COIN approach while London gradually left control of the territory to local militias. In fact, despite the increasing violence in 2005, London did not increase the British military presence in the province, while several NGOs and civilian agencies were forced (as occurred in Nasiriya) to reduce their reconstruction activities. Thus,

66 Website of the Air Force on Antica Babilonia, available at: http://www.aeronautica.difesa.it/Operazioni/Internazionali/Pagine/Iraq.aspx

67 *The Economist* (2009).

68 *The Economist* (2009).

in order to avoid widespread violence, the UK restrained its military presence on the ground (similarly to the Italians in the Dhi Qar province). The two above-mentioned operations (especially Charge of the Knights) thus enabled the British to emphasize the effective transition of responsibility to the local police and military forces, allowing the UK to leave the country and to concentrate all its efforts in Afghanistan.[69]

As reported by Marston (2009), Operation Sinbad, which was carried out in 2006 to 'contrast the local racket' in Basra, did not achieve any evident success in terms of 'clear, hold and build'. It was a costly mission for which the UK did not provide adequate resources. As highlighted by the British officer who mentored the UK commander in Basra, Richard Iron (2013), the timing of Sinbad (before the American surge) was essentially wrong and it was disconnected from the US strategy.

In 2007 the British patrols came constantly under attack by the local Shia militias. Thus the mediation with local militias for which Italy was severely criticized in 2004 appeared to be a good solution for the UK command as well. Consequently, the British forces moved from Basra City to the airport, leaving the local security forces in control. However, they were unable to accomplish task and, under US pressure, the UK troops were forced to regain stronger control of the city. Operation Charge of the Knights thus began in 2008: an intervention by Iraqi forces with the support of Western allies (for instance through drones). The operation 'cleared' Basra and created the conditions for a political agreement, and it also enabled the British Army to 'leave with honour'.

Both the UK and Italy were able to modify their approach and force structure in Afghanistan; but in Iraq such change proved much more complicated.[70] Two main aspects afford better understanding of the problems concerning the transformation of British forces in the Middle East.

The first is the problematic adoption of a proper COIN strategy. The lack of a comprehensive approach in Iraq fostered strategic ambiguity (Griffin, 2011). Apart from the political intent to avoid stronger (and more expensive) engagement in an unpopular conflict, other factors hindered the development of an effective counterinsurgency strategy in Iraq. For instance, the British COIN manual was not updated (so that the conceptual reference became the 'Petraeus doctrine') and the six-monthly rotation of troops was an obstacle to coherent plans of action.[71]

---

69 Prime Minister Gordon Brown announced the withdrawal of the remaining 4,100 British troops in December 2008.

70 France did not deploy its forces to Iraq. However, the lessons learned from the Middle East were also relevant for the French armed forces, especially regarding IED counter-warfare. See Coste (2009) and Girauld (2009).

71 It is worth noting that the 'Army's Stability Operations in Iraq (OP. TELIC 2–5): An Analysis from a Land Perspective, 2006' generally referred to insurgency and the need for a COIN approach on the ground. However, the *Army Field Manual, vol. 1, part 10, Countering Insurgency*, was published only at the end of 2009.

In a sense, the roles were reversed: whilst in the past US problems had been compared with Britain's record of successful COIN campaigns, FM 3–24 became the shared guideline for all the Western forces after December 2006 (Farrell and Gordon, 2009).

At the very beginning of the operation, when the violence was at its lowest level, the soft approach adopted by the British forces appeared to be successful. 'Early in 2004 British staff officers were tasked with persuading US commanders in Iraq to adopt a counterinsurgency strategy based on British counterinsurgency doctrine and experience from Malaya' (Egnell 2011, p. 299). But together with the minimum use of force, the UK troops implemented the peacekeeping principle of 'do no harm', refraining from population-control measures whereas COIN presumes the establishment of presence on and control of the territory (Ucko, 2010).

Hence when the difficulties began, the UK Ministry of Defence did not adjust its approach (Iron, 2013). Finally, as in Nasiriya, the problem of insurgency was wrongly characterized as criminality also in Basra (Iron, 2013).[72]

The second crucial aspect concerns the shortages that affected the British operation. While for Antica Babilonia the initial absence of proper equipment was due to the political interpretation given to the mission by Italian decision-makers, in the case of the UK, 'the pressures of Iraq and, especially, Afghanistan had exposed flaws in the equipment, organizational infrastructure and doctrine' (Griffin 2011, p. 318). The length and the demands of the simultaneous Iraq and Afghan campaigns were 'disproportionate' (Strachan 2011, p. 63). In sum, 'Britain's shortage of men and firepower, coupled with a different level of political commitment, goes some way to explaining why the British have been inclined to limit their ambitions'.[73]

The British inquiry on the conduct of the Iraq War (so-called Chilcot Inquiry 2009) revealed the evident shortcomings in Basra: a lack of troops (reduced from 46,000 in May 2003 to 10,500 in August), and equipment,[74] limited funding for development, and the absence of civilian experts on reconstruction. Oddly enough, in October 2003, Whitehall allocated no funds specifically to reconstruction.

In conclusion, while in Afghanistan the armed forces were able to correct the initial missteps, learning from mistakes, this did not happen in Iraq for either the Italian and British forces. Some modifications occurred (new equipment and efforts

---

72 As reported by Iron, Basra was presented as Palermo, not Beirut. In addition, the British initially believed that smuggling would help to kick-start the local economy. However, the militia controlled the smuggling (especially through harbours) significantly increasing its income in the 'period of accommodation'.

73 The Economist (2009).

74 However, thanks to the Urgent Operational Requirements, the British contingent received several items of new equipment, such as the Mastiff armoured vehicle. See Farrell and Gordin (2009).

in terms of training local forces[75]), but the unpopularity of the operation among Italian and British public opinion (Battistelli et al., 2012; Coticchia, 2014a; King, 2011) hindered stronger engagement in the mission (with constant adherence to the controversial soft and non-intrusive approach), it being preferred (mainly in the case of the UK) to give more resources and commitment in Afghanistan. The military experience of ISAF and OEF proved useful in the Middle East, where both PRTs and FOBs were created. In terms of emulation, both operations highlighted the primary role played by the US example (from the COIN doctrine to new digitalized equipment). Greater jointness and multinational interoperability confirmed this tendency. Also for this reason, it is worth examining the deployment of European armed forces, and their process of transformation, in a crisis where Washington was not present at all: Lebanon.

Lebanon: The 'Core' Mission?

Operation Leonte

The Italian military contribution made to the UN peacekeeping operation in Lebanon in 1982 was the most important mission undertaken by Italy during the Cold War.[76] It was the first intervention in which Italy deployed a large number of troops far from national borders. Despite its historical importance, the operation was an isolated event for Italian Cold War defence policy.[77] In fact, only after the constraints of the bipolar era disappeared did Italy start constantly to deploy soldiers abroad.

By contrast, the military role performed in 2006, as well as the rapid deployment of the national contingent, testified to the significant evolution of Italian defence policy in recent decades. After the Israel-Hezbollah war, Italy launched Operation Leonte with the purpose of reinforcing the United Nation Interim Force in Lebanon (UNIFIL). The goals of the UN intervention were to monitor the cease-fire, to support the Lebanese armed forces, and to guarantee security for the carrying out of reconstruction activities. Within the multinational contingent, which was now composed of almost 11,000 units, around 2,500 Italian soldiers were deployed to southern Lebanon to command the West Brigade of the UN force.

The Italian government conceived Leonte as a traditional peacekeeping mission, but contrary to what happened in the case of Antica Babilonia, the peace framework did not prevent decision-makers from providing an adequate force

75 For instance, the British military participated in the establishment of joint security stations throughout Basra (Iron, 2013).

76 For a comprehensive analysis of the approach adopted by Italian armed forces in 1982 see, among others: Angioni (1984).

77 For a detailed analysis see Calossi, Calugi and Coticchia (2013). Italy undertook two missions (in August-September 1982 and September 1982-March 1984).

structure for the intervention (Ignazi, Giacomello and Coticchia, 2012). In fact, the Italian contribution to the robust peacekeeping mission (Ronzitti, 2006) was based on the main assets of the armed forces, such as units for mine-clearing (dramatically necessary in Lebanon because of cluster bombs), CIMIC, NBC, Special Forces and *Carabinieri*. Also the Air Force and Navy played an important role, especially at the very beginning of the operation.[78]

The most controversial aspect of the intervention was the supposed disarmament of Hezbollah. Possible clashes with the Shia militia alarmed the public and heightened the political debate. Overall, despite some attacks against the Italian contingent and the enduring regional crisis, the operation was able to achieve compliance with the ceasefire and to support several development projects and programmes.

Apparently, UNIFIL was the 'core mission' for the transformed Italian post-Cold War armed forces: a 'traditional' peacekeeping operation that required joint and multilateral efforts, a comprehensive approach, a constant focus on reconstruction, digitalized capabilities, and power projection able rapidly to interpose troops between belligerents. Did the deployment of forces in southern Lebanon really mirror the key elements stressed by Italian doctrine, confirming them as vital dots in the national process of military transformation? Did it do so also on adopting a comparative perspective?

Mobility and power projection capability

As previously stated, Operation Leonte clearly demonstrated the evolution of Italian military operations abroad in terms of mobility: in a few days all the 'best' military means were made available to UNIFIL (Coticchia, 2013).

The mission is probably the best example of the constantly advocated readiness of forces, especially in regard to naval mobility. The amphibious projection was guaranteed on the first day of the action by the simultaneous availability of both the aircraft carrier *Garibaldi* and three San Giorgio/San Giusto class units (Cosentino, 2010). By means of these platforms Italy transported and supported an inter-force totally equipped 1,000-units contingent. An 'amphibious group' was specifically created to support the entry force in the initial phase of the stabilization. The severe test of Italian power projection capability was brilliantly passed.

Comprehensive approach

In terms of 'comprehensive approach', the test posed by the Lebanese scenario was not as complex as that of Afghanistan and Iraq, especially as regards the level of post-war violence (Della Longa and Giuli, 2008, Ignazi, Giacomello and Coticchia, 2012). As had occurred in the past, Italy adopted a non-intrusive and soft approach primarily focused on minimizing the impact of military presence and establishing a positive relationship with the local population. Moreover, dialogue with key Lebanese military and political actors (such as the Shia militias)

78 Italian forces were involved in the Maritime Task Force 448 – UNIFIL.

was part of the Italian 'traditional' stance (Di Feo, 2008). The Italian presence was generally perceived as positive.[79] As stated by the former Chief of Staff of the Army, Giuseppe Valotto, even when formal rules of engagement would have allowed the stronger use of force, the Italian battle groups in Lebanon preferred to deter instead of being involved in open clashes (which were rare).[80] UNIFIL thus illustrates the main components of the Italian 'comprehensive approach': civil-military relations, mine-clearing, and the training of local forces.

Differently from Afghanistan and contrary to what happened in Iraq, dozens of Italian NGOs were actively involved in the Italian efforts for reconstruction. The former commander of UNIFIL stressed how CIMIC units had benefited from such engagement to promote effective coordination among several actors and devote specific attention (and resources) to quick impact projects (Serra, 2012). The overall efforts made by Italian forces in local development and reconstruction (chiefly in the health and education sectors) were remarkable (Della Longa and Giuli, 2008).

The second important feature of Italy's multidimensional and 'holistic' approach within UNIFIL concerns mine-clearing. Italian forces had developed considerable skills in this field after years of military operations abroad (starting with Mozambique in the early 1990s). Both mine-clearing operations and mine-risk education (largely in regard to cluster bombs) became vital pillars for the intervention (Mason, 2007). Thousands of bombs and grenades were removed, 'clearing' several areas in southern Lebanon, consequently achieving the general goal of establishing trust with the local population (Vetere, 2010). Once again, the corps of engineers performed the prime function in the Italian operations.

Finally, as recognized by the former commander of UNIFIL General Graziano (2013), the training of local police and army officers was a crucial component of the national 'comprehensive' approach. The General directly referred to the McChrystal doctrine (which is based on mentoring, training and assistance) comparing it to the UNSC 1701 (the resolution that authorized UNIFIL II). Thus, as happened in Afghanistan and Iraq, increasing local responsibility for managing security became the priority.

Digitalization and force protection

Operation Leonte provided all the best national military assets on the ground. However, owing to the absence of a constant and dramatic level of challenges for the troops, the employment of net-centric warfare in order to guarantee force protection was limited to the initial months of the mission.[81] For instance, although the Pinerolo Brigade was the first unit that tested (in Afghanistan) the Freccia digitalised armoured vehicle on the ground, the forthcoming deployment of the

79 Author's discussion with a Hezbollah political leader in Haret Hreik, Southern Beirut (May 2011).

80 Quoted in Ciampini (2010).

81 Interview: Staff Officer, General Staff of the Army.

mechanized brigade in Lebanon (2015) will not involve the presence of the Freccia because of the different nature of the local context.[82]

However, the armed forces rapidly adapted their approach and equipment also in the case of UNIFIL. For example (largely after the Nasiriya attack) the system established to protect national compounds was entirely revised, also profiting from technologically advanced weapons to increase operational awareness.

Jointness and Multinational Interoperability

UNIFIL was a considerable innovation to the set-up of the chain of command of UN missions. Italy had suffered the consequences of ambiguity in the strategic planning and coordination of previous UN operations, such as UNSOM in Somalia (Loi, 2004). As reported by Hatto (2009), European countries were reluctant to contribute to UNIFIL in the absence of some special command and control arrangement. Rome applied pressure for the creation of a specific Strategic Military Cell (SMC) within the Department of Peacekeeping Operations (DPKO). The goal was to promote the more integrated coordination of activities on the ground, as well as to ensure the better strategic direction of field operations (Mason, 2007). The USA, Russia and all the nations have provided troops to UNIFIL were enclosed in the cell. This innovation was an important step towards further multinational interoperability for Italian defence. The harmonization of guidelines and procedures was essential in a broad multinational operation involving civilian personnel from 117 different countries and troops from 38. As stated by the former commander of UNIFIL, General Serra (2012), from the outset Italy required overall improvement in the effectiveness of the joint structure, especially because of the burden of its command responsibilities. The evolution of the Italian approach is here evident. Italy has traditionally considered the multilateral framework of the military mission abroad as a necessary premise for national engagement (Bonvicini et al., 2011). Values (Ratti, 2011), prestige (Davidson, 2011) and 'punching above weight' have been the main drivers of this attitude.

The case of Lebanon illustrates Italy's transformation as a security producer demanding not only a general multilateral structure but also innovative tools in order to improve the effectiveness of interoperability. And this is because of the mounting responsibility that Rome has assumed in the command of missions abroad, as during UNIFIL. As stated by Italian troops, the structure of UNIFIL did not provide a general command like that of ISAF, and this made the sharing of results more complicated.[83] The purpose of the innovation supported by Italy within the DPKO was to enhance

82 Interview: General Commander, Pinerolo Mechanized Brigade.

83 Interviews: Officer A, Folgore Parachute Brigade, Officer B, Folgore Parachute Brigade.

interoperability; however, the SMC allowed for more general coordination but it was not the UN command post.[84]

In summary, Italy's commitment testified once again to its evolution in terms of growing jointness and multinational interoperability, conceived as crucial operational and procedural requirements for the positive outcomes of a mission, rather than as general premises for the involvement of armed forces.

France in Lebanon

Together with Italy, France assumed the main political and military role in the Lebanese crisis. The French contingent, which had been deployed in Lebanon since the end of the 1970s, consisted of around 2,000 troops, a tank battalion (based on Leclercs), a mechanized infantry battalion (with AMX 30 AMX-10P armoured infantry fighting vehicles), a signal battalion, and an engineer regiment (Haddad, 2010). Indeed, also in the case of the French contingent, mine-clearance was a crucial task. Overall, heavy artillery weapons (such as AMX 30 AUF1) equipped the force.

Though far from the risky scenario of the Afghan Kapisa province, also southern Lebanon presented risks and challenges. The French troops were seriously attacked three times with roadside bombs (July and December 2011), fortunately without suffering any casualties.[85]

Two main features emerge on considering the French commitment to the UNIFIL operation. The first relates to the adaptation of equipment and approaches over the years in Lebanon. French forces strove to maintain a lower and softer profile on the ground (Haddad, 2010). Chapter 3 illustrated how the most recent *Livre Blanc* (2008) emphasized proper understanding of the local context as a precondition for success. As reported by the Joint Forces Centre for Concept Development, Doctrine and Experimentation, as well as by the Forces Employment Doctrine Centre, a positive relationship with the local population is a crucial factor in contemporary military operations.

The French contingent was adapted in the field in order to fit with the above-mentioned requirements. For instance, tracked vehicles and tanks like the Leclerc were seen by the population as 'much more aggressive than wheeled armoured vehicles', and they were consequently replaced with wheeled vehicles.[86] The new digitalised VBCI, which was deployed in 2010, also provided better protection against IEDs. Moreover, most of the vehicles were equipped with updated counter-IED jammers. Thus, the process of adaptation occurred but it took a considerable length of time because the local population reacted hostilely to

84 For a critical perspective on the new strategic cell in UNIFIL see Hatto (2009), according to whom the SMC cannot be conceived as a blueprint for a more generalized command and control structure for UN operations.

85 Army Recognition (2012). In the summer of 2010, a French convoy was ambushed in Tulin.

86 See Army Recognition (2012).

the 'aggressive' French military approach, constantly complaining about the noise caused by Leclercs in small villages (Coticchia 2013).

The second feature concerns jointness and multinational interoperability. On the one hand, the French proved to be properly prepared for involvement in all multinational command structures, and its joint forces performed an important function within the national contingent (which was a Joint Service Battlegroup). On the other hand, practical relationships with other allies at platoon level were hindered by the strict UN rules and procedures (Haddad, 2010). However, at the command and brigade levels, contacts and information sharing with Italians, Spaniards and others were considerable. In sum, the interoperability among allies, which was a constant feature of force transformation in post-2001 operations, was also apparent in the case of Lebanon, although the UN structure (notwithstanding the improvements brought by the Strategic Military Cell) was an obstacle to unified command.

Libya: The Problems of Re-adaptation

Italy in Libya

The military mission against the Libyan regime was officially undertaken in order to provide the 'necessary measures to protect civilians and civilian-population areas under attack or threat of attack in Libya' (UNSC Resolution 1973/2011). Thus, for the first time, the UN considered the so-called Responsibility to Protect – R2P (Patrick, 2011) as the main goal of the intervention, which was originally carried out by a multinational collation (Operation Odyssey Dawn) before NATO assumed the lead through Unified Protector.

Air strikes, the enforcement of a no-fly zone, the monitoring of an arms embargo, and 'humanitarian assistance' were the main tasks of the mission (Vira and Cordesman, 2011).[87] Italian foreign policy was initially ambivalent (Miranda, 2011) also because of Italy's close political and economic relations with the Gaddafi regime, especially in the fields of energy, infrastructure, military equipment, and the combating of illegal immigration and organized crime (Croci and Valigi, 2013). International pressure and the need to maintain an influential role in such a strategic country were crucial in convincing Italian government to revise previous positions (Coticchia, 2011; Lombardi, 2011; Davidson, 2013). Hence Rome finally made its contribution to the operation. The military commitment was extremely significant also in terms of the bases (seven) that Italy provided for allies. As stated by the official report of the Italian Air Force, national assets (such as Tornados,

87 In the meantime, the 'Hermes mission' was undertaken by Frontex to support the Italian authorities in the fight against illegal immigration.

F16 Falcons, Eurofighter 2000s, AMXs, Predator Bs, C-130s, etc.)[88] 'have carried out patrol, air defence, air refuelling, reconnaissance and neutralization of military targets missions. Overall, Air Force aircraft have carried out over 7% of the total missions over the Libyan skies.'[89]

Unified Protector (March-October 2011) was the largest Italian Air Force operation after WWII. In summary, Italian aircraft undertook 1,900 sorties, which represents the fourth military contribution after those of the USA, UK and France. But national engagement was not limited to the Air Force. According to Willet: 'Italy took a leading role in the NATO operation' (2011, p. 9). The entire naval component of Unified Protector was under Italian Command. According to the official figures, Italy deployed 16 warships (such as the carrier *Garibaldi* with AV-8B Harriers, amphibious assault ships, corvettes, destroyers and frigates), 3,500 soldiers, and two submarines.[90]

The operation finally toppled the Gaddafi regime, which collapsed in the autumn. Despite the 'success' of the mission (later contested by the country's dramatic instability),[91] two main problems are apparent if we consider the military operation in detail.

The first problem is burden sharing. Although the US Defence Guidance (2012) considered the Libyan intervention to be a positive example, analysis of the effective contribution made by the USA's allies reveals a different picture. The overall military contribution of the Europeans was considerable (Quintana, 2012). However, only six European countries (Belgium, Denmark, France, Italy, Norway, and the UK) contributed to strike missions. Other nations (such the Netherlands, Sweden and Turkey) deployed aircraft only to enforce the no-fly zone. NATO was divided, and important members (Germany above all) were reluctant to intervene. Less than half of NATO members were involved in the mission. 'The burden of the Libya operations is increasingly borne by a small number of countries. Others have erected strong political obstacles to participation, or caveats on their military contributions' (Cordesman and Vira, 2011, p. 7). Moreover, despite the crisis at its borders, the EU's contribution was almost non-existent (Coticchia, 2011).

The former US Defence Secretary, Robert Gates, linked the problem of burden sharing with the lack of adequate military capabilities. In other words, the allies

88 Tanker aircraft (KC-130J and KC-767A) also undertook air-refuelling missions. A G.222VS was employed for the detection of electromagnetic emissions.

89 Official report on the mission, available at:
http://www.aeronautica.difesa.it/Operazioni/Internazionali/Pagine/Libia(2011).aspx

90 The AV-8B Harriers deployed on the *Garibaldi* performed air strikes and NFZ enforcement missions (Cenciotti, 2011). The data on the Navy's contribution are taken from the official report of the Ministry of Defence, available at: http://www.difesa.it/Operazioni_Militari/operazioni-internazionali_concluse/Libia-odissea-alba/Pagine/Contributonazionale.aspx.

91 According to the ISS, (2011) NATO's air campaign was successful in its capacity to halt the counter-offensive of Gaddafi's forces, maintaining air dominance and supporting the rebels. For a critical perspective on the air campaign see Cenciotti (2011).

'Don't participate because they can't [...] It has become painfully clear that similar shortcomings, in capability and will, have the potential to jeopardize the alliance's ability to conduct an integrated, effective and sustained air-sea campaign'.[92]

Indeed, the second problem of Unified Protector consisted in the shortcomings of the military capabilities provided by European countries. Despite an active contribution at both diplomatic and military level, the recent literature has revealed several flaws: for instance, chronic shortfalls in ISTAR capabilities (intelligence, surveillance, target acquisition, reconnaissance) were widely reported (Cordesman and Vira, 2011; Quintana, 2011; Johnson and Mueen, 2012). Europeans depended closely on the support provided by American JSTARS (joint surveillance target attack radar systems) and AWACS (airborne warning and control systems) (ISS, 2011). Moreover, the US flew around 30 of the 40 air-refuelling tankers (ISS, 2011). Finally, bomb shortages were reported in the last weeks of the intervention (Cenciotti, 2011). According to Hodge: 'Without American intelligence, surveillance and reconnaissance (ISR), tanker, and suppression of enemy air defence (SEAD) capabilities, it is unlikely that the European member-states could have sustained a protracted air campaign' (2013, p. 362). In some cases, the multinational contingent effectively modified its structure, providing additional equipment made available by the evolution of the military context on the ground. For instance, unmanned aerial vehicles (US and Italian Predators and French Harfangs) were uplifted in August, delivering better information in a crucial phase of the war (Quintana, 2012).

Looking at the Italian military commitment, the operation in Libya makes it possible to assess the most recent transformation of Italian armed forces in a more traditional warfare context, far from the peacekeeping or counterinsurgency missions regularly undertaken in the new century.

Mobility

Operation Odyssey Dawn in Libya illustrated the enhanced Italian capabilities in terms of amphibious projection capability and strategic mobility, and especially the role of the Navy, which evacuated civilians (mainly from Benghazi) and enforced the embargo with its main units. Rescue operations are the best example of the constant need for increased mobility as conceived by national strategic documents.

Despite initial political ambiguity and vacillations (Miranda, 2011), the Italian armed forces were able to provide a rapid response, giving their relevant military contribution, essentially through the Navy and Air Force.

Comprehensive approach

Unlike in previous conflicts, the Italian Army did not play the leading role. However, both during the war and after the end of the multinational intervention, Italy deployed units for the mentoring and training of local forces (e.g. Operation Cyrene). Therefore, the Italian focus on training once again emerged as a prime

---

92 Reported by *The Wall Street Journal* (2011).

asset in re-establishing adequate security in a post-conflict scenario. This aspect represents an evident continuity in the Italian approach consistent with the national doctrine. The same applies to the holistic perspective on security that characterizes Italian strategic papers, and it was visibly implemented in the Libyan operation. Consistently with the multidimensional view of contemporary security challenges, Italy deployed (and employed) its armed forces (the Navy) to deal with non-military treats such as smuggling and illegal migration (Ceccorulli and Labanca, 2014). As occurred in the 1990s in the Adriatic Sea (during the operations Alba in 1997 and Allied Harbour in 1999), Italian warships played a leading role in combatting the trafficking of arms, drugs and persons consequential on military and humanitarian crises. Therefore, Unified Protector exhibits the concrete application of the multidimensional strategic perspective on the nature of current threats to national security. Subsequent military operations in the same area (such as Mare Nostrum[93]) have substantially confirmed this view, although at the same time they have provoked heated debate on the concrete effectiveness of armed forces in contrasting non-military challenges and possible gaps between aims (even humanitarian) and means.[94]

Digitalization and force protection

Among patrolling, air defence, attacks on predetermined ground targets, surveillance and reconnaissance, air refuelling and detection of electromagnetic emissions, the suppression of enemy air defences (SEAD) was the capability of the Italian Air Force most appreciated by the allies. In fact, only Italian warplanes (the Tornado ECR – Electronic Combat Reconnaissance), together with those of the Americans, were equipped on board with this capability during Unified Protector. Moreover, as reported by the Air Force, the Tornado IDS (Interdiction/Strike aircraft), the AMX aircraft and unmanned Predator B, 'carried out extensive reconnaissance activity and the acquisition of aerial images and valuable data for the conduct of operations (ISR – Intelligence Surveillance Reconnaissance)'.[95] Italian jets achieved great accuracy in air strikes (over 96% according to official figures), doing so also with laser guided precision weapons. Moreover, the Air Force used long-range Storm Shadow cruise missiles for the first time. The campaign thus testifies to the implementation of new or modernized assets by the Italian armed forces, especially by the Aeronautica. After years of PSOs and COIN missions, Italy

93 On 'Mare Nostrum' see: http://www.difesa.it/OperazioniMilitari/NazionaliInCorso/MareNostrum/Pagine/default.aspx.

94 The 'Mare Nostrum' operation has been criticized for a lack of clarity in its strategic aims: should it halt the flow of migrants or should it help the refugees? For a general perspective see, among others Del Panta (2014).

95 See the official account of the mission available at: http://www.aeronautica.difesa.it/Operazioni/Internazionali/Pagine/Libia(2011)_eng.aspx.

demonstrated itself able to make a significant military contribution also in a more traditional operation. It did so consistently with the recommendations of strategic documents, in particular those concerning the constant need to increase the digitalization of forces.

However, as mentioned above, Unified Protector also exhibited shortages and deficiencies, such as its dependence on the US ISTAR capabilities. ISTAR was not 'prioritized'[96] by Italy and the other European countries. Moreover, the campaign also demonstrated limitations in terms of intelligence, given the scant amount of information available on the ground (Johnson and Mueen, 2012).

In the case of Italian forces, problems arose in regard to exhausted fuel, resources for maintenance, and 'smart' ammunitions (Quintana, 2012). In addition, the Navy used the *Garibaldi* and the frigates, while the brand-new carrier *Cavour* played no combat role because it still lacked its air wing (it was still awaiting the controversial F-35).

Jointness and multinational interoperability

Within the multilateral framework of Unified Protector, the Italian military assumed vital responsibilities in the chain of command, both in the targeting division and the Standing MNATO Maritime Group 2. Italy hosted the Joint Force Command (JFC) in Naples while concurrently contributing at tactical level to the Combined Air Operation Centre 5 (CAOC 5). As reported by interviews, the mission in Libya well demonstrates the myth of the Air Force's 'autonomy' because multinational interoperability currently represented a vital requisite for all contemporary military interventions abroad.[97]

However, according to some authors, the level of integration during the initial operation in Libya (Odyssey Dawn) was not entirely satisfactory,[98] as testified by the fact that each country had to have its own tankers (Cenciotti, 2011). Things changed after NATO took over, and the Italian Air Force was adequately integrated into the multilateral context. The targeting division provides a very good example of the joint process. As stated by the Air Force: 'All military objectives assigned to Italian aircraft by NATO were prescreened by a general officer in the NATO Chain of Command – the technical term, a red card holder – to verify compliance with the guidelines established by the policy authority'.[99] According to the Armed Forces, the engagement of a very large percentage of the military targets assigned, while avoiding 'collateral damage', was closely connected to this integrated process.

96 Interview: Taylor.

97 Interviews: Officer A, Italian Air Force, Institute of Military Aeronautical Sciences, Officer B, Italian Air Force, Institute of Military Aeronautical Sciences, Officer C, Italian Air Force, Institute of Military Aeronautical Sciences.

98 Cenciotti (2011) criticized also the strategic direction of the mission, stressing that: 'It took 2 months to understand that it was better to start targeting Gaddafi's capacity to resupply his forces on the front rather than attacking each single vehicle on the frontline'.

99 *Rivista Aeronautica* (2012).

The most important aspect of the role performed by the NATO framework is probably the way in which interoperability and multilateralism have been conceived by the Italian military. In other words, the multilateral value acquires a sort of 'procedural dimension': 'Every nation is structured to make war in its own way … France, Great Britain … we are not. The others are able, unlike Italy, to set up operations alone, so they can provide 'closed packages', we are not'.[100] Thus interoperability is a mandatory feature, an essential premise for Italian engagement, not only because multilateralism is a transitional guideline of Italian foreign policy (as well as a fundamental requisite for approval by public opinion) but also, and especially, because of its added value in terms of shared training, procedures and approaches. Thus Italian pilots greatly preferred a NATO operation to a (less integrated) multinational mission.

France and the UK in Libya

The UK Operation ELLAMY made a considerable military contribution to the multinational mission in Libya, with 4,000 personnel, 37 aircrafts, 16 ships, submarines and auxiliary assets assigned to the campaign.[101] The official report declared that military performance was generally good, and it emphasized the important role played by RAF Typhoons, and AH-1 Apache helicopters. From the perspective of the RAF, the mission made it possible to shift from air support to targeting operations. As stated by Major Mick Neville, Officer Commanding 4 Regiment AAC: 'In Afghanistan we are used in a reactionary way, giving support to guys on the ground, so it was nice to be involved in deliberate targeted operations'.[102] Unified Protector was based on the use of full-spectrum military capabilities, rather than on COIN activities supported with air and maritime services (Quintana, 2012). For example, maritime forces were involved across the entire spectrum of conflict, from non-combatant evacuation operations to conventional operations with Tomahawk cruise-missile strikes (Willett, 2012).

Therefore, in a new scenario, after years of counterinsurgency, the challenges and the operational requirements significantly changed. In order to assess the UK capability properly, we should understand how and to what extent the approach (as well as financial constraints) promoted by the *Strategic Defence and Security Review* – SDSR (which was published just a few months before the operation) affected the UK's military capabilities during the intervention.

According to Cornish and Dorman, the operation in Libya 'was broadly consistent with the MoD's planning assumptions' (2012, p. 222). As Willett (2012) writes: 'Libya was the first operation in which the Royal Navy's newly

---

100 *Rivista Aeronautica* (2012, p. 25).

101 See Ministry of Defence (2012). The UK, as well as France and (probably) Italy, also provided special forces on the ground. Their role was crucial in supporting the rebels before the assault on Tripoli in August. See, among others, Kometer and Wright (2013).

102 See Ministry of Defence (2012).

established Response Force Task Group (RFTG) was deployed. A post-SDSR amalgamation of the UK's Carrier Strike and Amphibious Task Groups, the RFTG was one of the UK's most significant contributions, and also enabled the Royal Navy to test a number of new operational concepts, including air/sea integration' (2012, pp. 43–4)

Clarke (2012) acknowledges the general success of the operation, but he also emphasizes the inconsistency between the Libyan operation and the recent National Security strategy, which did not rate the Libyan crisis (in terms of risks and national interests involved) as strategically important. In other words, the so-called 'NSS system' (e.g. the checklist of priorities designed through the risk management approach adopted by the Strategy) would not have 'suggested' a military intervention in Libya, given the risks and interests involved.

More specifically, as stated by Johnson and Mueen (2012), both the Air Force and Navy had to divert assets from other tasks to carry out the mission. For instance, owing to the high usage rate of Tornados and Typhoons, the UK lacked adequate air defence at home, while the cabinet also reduced the naval contribution to counter-piracy operations (Quintana, 2012).

The above-mentioned inadequacy regarding ISTAR was certainly due to recent cuts that forced the UK to extend the service of the Nimrod R1 reconnaissance aircraft (Coticchia, 2011). Overall, the UK 'too had some issues with supply, although principally with the hugely successful Dual Mode Brimstone weapon that had been developed in low numbers as an Urgent Operational Requirement for Afghanistan' (Quintana, 2012, p. 35). There were also limited numbers of tankers for the crucial air refuelling activities.[103]

Finally, London was unable to deploy an aircraft carrier, while the amphibious assault ship HMS Ocean acted as a 'helicopter carrier' (RUSI, 2011), supporting the allies in launching aircraft during OUP.

At the same time, this lack was the first test for the new UK-France bilateral defence cooperation arrangement. Probably, as illustrated by Cameron (2012), the war came too soon for adequate evaluation of the defence agreements. Indeed, some miscommunication occurred, so that after the end of the intervention the allies decided to prioritize 'joint work in the key areas of command and control; information systems; intelligence; surveillance; targeting and reconnaissance; and precision munitions'.[104]

In terms of interoperability, the involvement of several national assets without specific experience of coordination (especially of maritime operations) created inevitable problems (Willett 2012). Moreover, NATO relies on individual nations' intelligence. Consequently, the effective sharing of information without a

---

103 This forced to a considerable amount of fuel transferred per day, causing problems to the local Italian Infrastructure (Quintana 2012, p. 36).

104 See: 'UK-France declaration on security and defence', available at http://www.number10.gov.uk/news/uk-france-declaration-security. 17 February 2012.

centralized body proved difficult, also because the French were not in the NATO intelligence cycle (Quintana, 2012).

France made substantial efforts to promote better cooperation with allies, although they were mainly addressed to the UK. However it is worth noting that the Libyan operation was France's first intervention since it had rejoined NATO's integrated military command structure.

France assumed leadership of the mission, thanks to a very active diplomatic and military role. The national contribution (Operation Harmattan) was based on 4,200 personnel, 40 aircraft, 20 helicopters and the *Charles de Gaulle* carrier.[105]

French helicopters deserve special attention because they represented the key asset of the entire operation in line with the mobility advocated by national strategic documents. Indeed, SA-342 Gazelles armed with HOT-ATGMs, Eurocopter Tigers and Pumas carried out almost 90%[106] of the attacks launched with helicopters during Unified Protector. According to Cenciotti (2011), the French choppers, which flew within strike packages of 2–6 in cooperation with maritime gunfire support, were more effective than the British helicopters.

The GAM (*groupement aéromobile*) played a crucial role mainly in urban areas, engaging enemies more effectively and with fewer risks of civilian casualties (Coticchia, 2011). The need to avoid 'collateral damage' was prioritized over mere force protection, because helicopters were more vulnerable than aircraft to ground fire (Vira and Cordesman, 2011).

In summary, the French intervention was perceived as consistent with the strategic orientations emphasized by the 2008 White Paper, which focused on the regional instability in Sahel-North Africa. In addition, the mission in Libya was the first NATO operation undertaken by France after it re-joined NATO's integrated military command structure. Thanks to its active political and military role during the crisis, France reaffirmed its influence while maintaining operational discretion over its assets (Cameron, 2012).

In conclusion, Libya was a tough test for European countries, which had to re-adapt approaches and equipment largely shaped by a decade of peace enforcement, nation-building and, above all, counterinsurgency missions. Unified Protector illustrates the persisting significant capabilities of European countries in undertaking a more 'traditional operation', despite some shortages and with different degrees of consistency with recent national doctrines.

---

105 See the offical figures: 'Libye : point de situation n° 50 – bilan de l'opération Unified Protector', available at: http://www.defence.gouv.fr/operations/autres-operations/operation-harmattan-libye/actualites/libye-point-de-situation-n-50-bilan-de-l-operation-unified-protector

106 Libye : point de situation n° 50 – bilan de l'opération 'Unified Protector'.

Summary

The Italian units deployed in military interventions faced new types of challenges after the end of the Cold War, and especially in the post 9/11 era. Also for this reason, the demand for innovation has been largely perceived during, and after, the commitment of armed forces abroad. According to Lieutenant General Claudio Graziano, Chief of Staff of the Italian Army, the operational dimension is the 'centre of gravity of transformation' (2013, p. 10).

The empirical analysis conducted by this chapter described the evolution of the 'translation process' from the political to the military level, and it emphasized different patterns of adaptation on the ground.

Digitalization has been gradually enhanced while the growing 'power projection capabilities' have clearly distinguished the expeditionary nature of the forces, as well illustrated by the case of Lebanon. The interoperability with allies, especially through joint operations on the ground (mostly in Afghanistan), represents an indispensable element to understand the whole transformation of the Italian military. NATO has played a very significant role, both as preferred source of doctrinal innovation and as broad framework of emulation and 'learning by doing together'. The operation 'Unified Protector' in Libya reveals the vital importance of the multilateral context for Italian forces, primarily form a military point of view.

At the same time, the adaptation from a 'low profile approach' to a more proper stance (mainly in terms of equipment, rules of engagement or caveat) has been sometimes 'imposed' by the emerging underestimated threats, with severe consequences (as occurred in Iraq).

The next chapter examines the main findings of analysis of the coherence among deployment, doctrine and budget, also taking account of insights from theoretical analyses of military transformation. Assessing the main lessons learned and the interplay among different dimensions, the chapter illustrates patterns of adaptation, evolution and innovation of the Italian armed forces from a comparative perspective.

Chapter 5
Patterns of Learning and Change

In this chapter, we examine three broad dimensions of military innovation. First, we look at how doctrine 'performed' in the field and the amount of adaptation that took place in Italy's main post-2001 military operations. From growing multinational interoperability and digitalization to counter-IED strategies, analysis of Italian operations in Afghanistan, Iraq, Lebanon and Libya allows assessment to be made of the degree of coherence between the key elements of national doctrine and the evolution of deployment. Second, we observe how operational experience related to institutional learning, and how the Italian armed forces changed in response to field experience. The focus will be on the set-up of a process to institutionalize lesson-learning. Third, we assess the extent to which the interplay between operational experience and resources affected force transformation with specific regard to military assets. The fourth section summarizes some of the key features of learning and change in the UK and France.

From Planning to the Field

The literature review (Chapter 2) has already illustrated the need to pay closer attention to the link between bottom-up practices in military institutions and innovation, particularly with reference to learning and adaptation (Grissom, 2006). Therefore, the role of doctrine at the operational level should be investigated in detail, and the consistency between strategic thought and force deployment should be carefully assessed.

Chapter 4 provided the information required for proper evaluation of the degree of coherence between mission requirements and the key elements of Italian strategic documents. It showed both cases that featured a significant correspondence and situations in which it has been extremely difficult to move from doctrine to the field. Overall, the main findings are as follows.

First, according to Italian doctrine, the armed forces should be essentially 'expeditionary' in their nature. In terms of mobility and power projection capability, the evolution of Italian defence has been considerable. Whilst at the beginning of the century, and specifically during the first weeks/months of Operation Enduring Freedom, some logistical setbacks occurred (especially for the Air Force), subsequent missions reveal a rather different picture. Already Nibbio demonstrated a capacity for the rapid deployment of transport over long distances, as well as the very first air assault carried out at battalion-level.

In addition, strategic documents frequently focus on maritime power projection, for instance through so-called coalition sea-basing. The best assets of the Italian navy (the *Garibaldi* carrier and several assault ships) were engaged in OEF as well. But it was Operation Leonte in Lebanon that provides a better example of the recommended readiness of forces, also thanks to the creation of an 'amphibious group' to support the intervention.· Specifically, amphibious capabilities represent the constant feature of Italian power projection through naval mobility, as demonstrated also by the recent rescue operations undertaken in the Libyan crisis. Therefore, Italy's principal post-2001 military actions exhibit overall improvements in power projection and mobility, manifold innovations at tactical level (such as the ground assault convoy), and substantial consistency with the doctrine concerning the expeditionary nature of the Italian forces.

Second, all the operations showed some recurrent elements of the 'national comprehensive approach'. Italian interventions are generally distinguished by two main actions: reconstruction and training. As repeatedly stressed in all the White Papers, the focus on reconstruction predominates in the Italian approach to the crises. PRTs or CIMIC units were the traditional means with which Italy had previously faced challenges in the field of aid and nation-building (for instance in the Balkans in the 1990s). The military operations reveal diverse patterns in the complicated nexus between development and security: in Lebanon, a traditional activity of Italian forces like mine-clearing was crucial for acquiring the trust of the local population. By contrast, the inability to improve security in Iraq inevitably affected the efforts to foster development. In fact, despite considerable attention to reconstruction (even if without NGOs and other independent civilian actors), the projects implemented had limited impact in Nasiriya because Italy did not adjust its soft and non-intrusive approach. During Antica Babilonia the contingent would have needed a proper counterinsurgency strategy (one, for instance, which forcefully pursued closer control of the territory). But, as evidenced by the conceptual gaps in the (limited) national strategic public debate, even the term 'COIN' was never taken into account, and the superficial 'peace mission' perspective was never abandoned.

Afghanistan well illustrates the evolution of the national approach, at least on the ground. Regarding 'the Italian way to COIN', we emphasize that Italian forces sought to adapt the traditional 'holistic approach' previously adopted in several missions (and based on training, reconstruction, development, and civil-military relations) to the local combat scenario. The main conceptual reference for Italian soldiers was constituted by the Field-Manual 3–24 (and by its 'implementation' of Petraeus and McChrystal). As said, very limited strategic reflection occurred in Italy on COIN, despite the significant experiences collected in interventions abroad (e.g. Somalia).[1] Interviews with Italian troops show the centrality of

1 A possible exception is represented by: General Staff of the Army, 2008. *Le Operazioni Contro-insurrezionali*, SME III RIF-COE.

NATO's conceptual references.[2] On the one hand, there is continuity in how Italian soldiers have focused on the relationship with the local population, devoting their primary efforts to economic and social reconstruction; on the other hand, this 'predisposition' was 'systematized' after 2008–2009 by the new surge strategy developed at US and NATO level.[3] The term 'counterinsurgency' was finally introduced into the Italian vocabulary, despite a specific COIN doctrine remaining almost absent. Therefore, in the case of ISAF the innovation was not related to the broad Italian approach, which remained 'comprehensive' but pertained to a more structured operational design that emphasized new elements for the Italian military.

The second main aspect of the Italian comprehensive approach is the training of local security forces. Also thanks to national assets, such as the *Carabinieri*, Italy has become a much-appreciated ally in contemporary missions. In Afghanistan an evident process of adaptation to the combat scenario took place even for the training, as illustrated by the increasing military dimension in renewed curricula. Besides training, also the mentoring of police and military forces became a vital characteristic of national operations, as evidenced by all the interventions considered (e.g. through the OMTLs).

In sum, the doctrinal centrality of the comprehensive approach is widely confirmed by military engagement. Especially after the problems encountered by Antica Babilonia, deployment in dangerous operations (such as Afghanistan) required more rapid transformation of the national approach, adapting and modifying previous traditional predispositions.

Third, doctrine constantly focuses on the growing needs of digitalization. Technological edge is viewed as crucial for contemporary military operations abroad. The role of brand-new equipment, especially with net-centric capabilities, has significantly increased: drones or integrated systems of command and control are now fundamental assets for troops.

Afghanistan, for instance, was a crucial test also in terms of net-centric warfare: new technologically advanced systems, such as the VBLM 8x8 Freccia with the C2 SICCONA systems were employed on the ground for the first time. In Libya, Italy provided crucial SEAD capabilities with Tornados. Unmanned aerial vehicles, such as the Predator, became constant components of the Italian contingent, especially because of their added value in terms of information awareness. Despite improvements, however, some shortcomings have been reported, such as the close dependence on the American ISTAR capabilities during Operation Unified Protector. The intervention in Libya highlighted two main findings pointing to this regard: on the one hand, Italian forces are perfectly

2 Interviews: Officer A, Folgore Parachute Brigade, Officer B, Folgore Parachute Brigade, Officer, NATO Rapid Deployable Corps Italy, Officer A, Taurinense Alpine Brigade.

3 Interviews: Officer A, Folgore Parachute Brigade, Officer B, Folgore Parachute Brigade.

able to provide high-technological engagement (e.g. for the suppression of enemy air defences), on the other hand, the Libyan operation evidenced limitations in carrying out a prolonged traditional action after years of low-intensity conflicts (e.g. shortages of ammunition).

Fourth, the adoption of digitalized systems relates to another recurrent requirement stressed by the doctrine: the safety of troops. Force protection has been prioritized at both doctrinal and operational levels. Although the new COIN strategy required the diverse engagement of troops on the ground, with more risks taken in patrolling outside the new armoured vehicles,[4] updated technological equipment was better able to protect contingents. Therefore, both digitalization and the comprehensive approach have gradually evolved in response to the threats of contemporary crises. 'Counter-IED warfare' testifies to such continuous efforts to improve protection against the most effective tactics developed by insurgents in Iraq or Afghanistan (and on some occasions in Lebanon). For instance, the 'Weapons Intelligence Teams' and the increasing focus on lessons learned collected after every attack are probably the best examples of this determination. But adaptation and innovation have not been limited to equipment like electronic countermeasures (jammers) or more adequate armoured vehicles (such as the Lince). Also approaches have changed in response to the operational reality. For instance, the engineer corps has evolved greatly in doctrinal terms, changing from a static approach to a dynamic one in order to eliminate the sources of threats. The Army established of the C-IED Centre of Excellence exactly with the aim to develop both appropriate doctrine at tactical level and training. By contrast, the systems adopted for the protection of Italian bases abroad have been updated only *after* the dramatic attack of Nasiriya. Nevertheless, the lessons learned from that tragic event are evident on looking at the well-equipped compounds of UNIFIL.

Fifth, in line with strategic documents, as well as with the traditional guidelines of Italian defence and foreign policy, jointness and multinational interoperability with allies distinguished all the interventions examined.[5] The operations heightened the degree of jointness among the Italian armed forces at tactical and strategic level. This requirement – already considered a precondition for commitment by the national and international doctrines – deeply affected the Italian military engagement, improving joint capabilities especially in regard to command and control and targeting. Military interventions imparted the main impetus for enhancement of multinational interoperability through shared military actions (also intense combat missions like those in Afghanistan), constant information sharing, and mutual procedures. The analysis of deployment has clearly demonstrated

4 Interviews: Officer A, Folgore Parachute Brigade, Officer B, Folgore Parachute Brigade.

5 Whilst the degree of interoperability with allies was the preeminent feature of the Italian mission, effective jointness among national services is more complicated to assess, also because of the major role played by the Army on the ground. For positive examples of integration among task forces in Afghanistan, see Ciampini (2010).

Italy's activism within a multilateral framework, as exemplified by its pressing for a new and more effective strategic cell within the United Nations Department of Peace-Keeping Operations (DPKO) before UNIFIL II. The coherence with doctrine was indubitably close, but while strategic documents highlight the importance of coordination with allies and within a multilateral framework, the Italian forces visibly manifested a preference for NATO's intervention rather than UN, European or multinational missions.[6] The possibility to have an integrated and effective command hierarchical structure (as in ISAF), and the amount of training with NATO's standards and procedures are the main reasons for this preference.

Finally, the analysis of deployment allows assessment of the overall Italian capacity to modify the approach on the ground as established by decision-makers, also in conformity with doctrinal references and budget considerations. To be noted is that the financial constraints of the Italian defence budget did not prevent the national contingent from having the weapons that it required.[7] This was the case of Afghanistan after 2006 (and even more so after 2008). All the soldiers that we interviewed acknowledged the efforts made by Rome to provide proper equipment (armoured vehicles, UAVs, helicopters, etc.) to face the menaces posed by the Taliban. On the contrary, in 2003, Italy had old military apparatus not entirely suited to the local combat scenario (e.g. the VM90). Thus the deployment brought a necessary renovation that combined with a different political approach in terms of caveats or rules of engagement (from a light force like Nibbio to a completely different operation). And the 'humanitarian' approach adopted in the case of Antica Babilonia, which was interpreted (and consequently designed) as a humanitarian mission, was mainly responsible for the inadequacy of the force structure. The Italian strategic documents focused on the need to prevent mission creep and to deal with possible asymmetric threats, configuring the contingent with advanced combat capabilities. This was not the case of the Iraqi mission, which never modified its unsuccessful approach (despite providing more adequate equipment after violence erupted). The strategy adopted by the Italian cabinet was not affected by the new Petraeus COIN, because the operation ended in 2006.

While in the case of UNIFIL no specific problems arose because of the 'adequate' composition of the contingent for a robust version peacekeeping, the carrier *Cavour* was not employed during the Libyan crisis (when the overall Italian naval contribution was extremely significant) because the F-35 aircraft had not yet been available.

In sum, doctrine interacts with operational experience in a positive way. As illustrated above, despite setbacks and problems, the overall coherence between strategic documents and deployment is relatively high as regards mobility,

6 Interviews: General Commander, Folgore Parachute Brigade, General Commander, Pinerolo Mechanized Brigade, General Commander, Taurinense Alpine Brigade.

7 Interviews: General Commander, Folgore Parachute Brigade, General Commander, Taurinense Alpine Brigade.

comprehensive approach, multinational interoperability and digitalization. The dynamic interplay between the dimensions of transformation reveals the centrality of field experience. Insights from the ground also promoted operational learning, which was often shared, systematized and disseminated within a multinational framework such as NATO. In addition, the lessons collected in military operations abroad paved the way for adaptation at institutional level.

Institutional Setup and Learning

The Institutions of Learning

Feedback from operational experience has been central in promoting transformation at the institutional level and in structuring organizational learning. For armed forces that have been to a large extent crystallized in the Cold War, the intense deployment that began in the 1990s inevitably acted as a key driver of change. This applies to all NATO's European countries, and all the more so to Italy because of its previous limited experience. The Gulf War and the Balkans started a process of change that involved several aspects of military institutions and the everyday lives of soldiers, primarily the need to coordinate with partners on an almost daily basis and – at the individual level – to overcome cultural barriers such as those caused by language (Soeters and Manigart, 2008). One of the simplest, but most effective, indicators of transformation in the recent past is the adoption of English as the working language of European militaries in their everyday routines. Assessing how the strong stimulus provided by experience was translated into transformation, and how that transformation was actually responsive to and coherent with the stimulus received, is a complex task. In what follows, we provide a preliminary analysis focused on how the Italian armed forces have modified their institutional structures tasked with institutional learning.

From an institutional point of view, major changes have consisted in the creation of structures (or the development of existing ones) to deal with knowledge retention and transfer: that is, with the cycle of lessons learned. The institutionalization of lessons learned is a recent process that has generated a new 'architecture' largely modelled on NATO.[8] Different levels of action, and therefore of analysis, exist. At the joint level, the *Centro Innovazione Difesa* (CID, Centre for Defence Innovation,) was created in 2009 within the *Stato Maggiore Difesa* (*III Reparto*) to develop and update doctrine. The Centre is an important innovation because it represents the first joint body to elaborate doctrine, and it constitutes the apex of the chain with reference to the entire process of lessons learned.

Together with new structures, the idea of jointness acquired increasing importance. With interoperability playing a major role in most operations, the

8 Interviews: Staff officer A, General Staff of the Army,Staff officer B, General Staff of the Army.

mechanisms and institutions dealing with jointness played an increasing role. The *Comando Operativo di Vertice Interforze* (COI – Joint Operational Command), which was created in 1997 (before the military intervention in Albania) constitutes the framework in which operations have been managed. The centrality of COI extends beyond strictly operation-related tasks because the Command is responsible for issuing directives on lessons learned and thus start the sharing process (see below).

In addition, it must be noticed how the concept of operating jointly has not been limited to the national level. In fact, some projects involving multinational units have recently emerged. With reference to Italy, the most important is arguably the Non-Standing Bi-National Brigade Command (NSBNBC) that connects the Taurinense Brigade and the French 27th Brigade de Infanterie de Montagne.[9] The technical agreement signed in 2011 identifies joint employment within a multilateral framework, information sharing and the development of standard operating procedures as the main goals of the cooperation.[10]

Although multinational interoperability is central, it is nonetheless not surprising that the Army, most heavily involved in recent operations – at least in Afghanistan – is at the forefront of the effort to improve the lessons-learned cycle. Three institutions play a specific role. The first is the *Comando per la Formazione, Specializzazione e la Dottrina dell'Esercito* (COMFORDOT, Army Command for Training and Doctrine) created in 2013 in Rome.

The second is the *Centro Simulazione e Validazione dell'Esercito* (CE.SI.VA, Centre for Simulation and Validation) of Civitavecchia. Previously a more traditional advanced training centre, the CE.SI.VA has been restructured with the mission of preparing officers to assume command posts in the field (at the brigade and regiment level) through training at Combat Training Centres designed to create realistic simulations of operational experience. The Centre also develops the cycle of lessons learned in the fields of training and operations, in coordination with the COMFORDOT, and accordingly produces doctrine down to the tactical level.

The third is the *Dipartimento per la Trasformazione Terrestre* (DTT, Department for Land Transformation), created in 2006 to provide a holistic approach incorporating Doctrine, Organization, Training, Material, Leadership, Personnel, Facilities – Interoperability (so-called DOTMLFPI) in accordance with NATO doctrine. This is strictly linked with the idea of NEC developed above.[11]

---

9 Another relevant multinational project is the cooperation between Julia Alpine Brigade and counterparts from Slovenia and Hungary.

10 The Joint Brigade will be operational in 2015 in Lebanon. The Taurinense Alpine Brigade was selected because of its extensive operational experience, especially within the multinational UNIFIL framework. Interview: General Commander, Taurinense Alpine Brigade.

11 New inter-forces 'packages' were established, such as the abovementioned C-IED centre or the NBC school at Rieti, as well as a new hub for the Predator A and B, the *Centro di Eccellenza per Aeromobili a Pilotaggio Remoto* (Centre of excellence for remotely-piloted

Key changes in the lessons-learned cycle have also occurred in regard to the link between learning and the transformation of the armed forces.[12] The process is based on the adoption of a system for the evaluation of the effects achieved and of the actions undertaken during operations and training in order to verify whether they reach the desired *end state*, and if they do not, to devise appropriate remedies (SMD, 2013). Measures of effectiveness (MoE) and Measures of Performance (MoP) are adopted to assess respectively the achievement of the desired effects and the outcome of actions, following the effect-based approach (incorporating the logic of Effects-based Operations, EBO). Force innovation comes about through a capabilities adaptation (*adattamento capacitivo*) intended to remodel the armed forces in the short, medium and long run. It arises from experiences in operational theatres and training activities through the constant restructuring of doctrine, assets and structures.[13]

The Lid/LL process is also a key starting point for the *adattamento capacitivo*. The specific authorities in charge of the process are the SMD III Reparto/CID for the strategic level, the COI for the operational level, and the *Comando Operativo Forze Speciali* (COFS, Operational Command for Special Forces, instituted in 2004) for Special Forces operations.

The revised cycle of lessons learned consists of five major phases.[14]

1. The *gathering phase* is based on the signalling by officers and soldiers of perceived problems in their activities (i.e. operations, training, and command) that 'cannot be solved at their own level' of competence and thus requires to move upwards in the chain of command so as to improve the 'effectiveness and efficiency' of the activities performed by the military across the multi-dimensional spectrum of DOTMLFPI. The agent that drafts this first document should be able to establish a cause-effect mechanism underlying the problem reported.
2. The *analysis phase* is conducted by expert officers in the relevant branches under the supervision of an officer responsible for Lessons Learned. In this phase, the objective is 'identify a line of action (…) to solve the problem encountered definitively'. The process involves a first phase of

aerial vehicles) of Amendola. Interviews: Officer A, Italian Air Force, Institute of Military Aeronautical Sciences, Officer B, Italian Air Force, Institute of Military Aeronautical Sciences, Officer C, Italian Air Force, Institute of Military Aeronautical Sciences.

12 Transformation is defined here as: 'a continuous process of elaboration, review, implantation and integration of concepts, doctrines and innovative capacities, activated in order to maximize the effectiveness and the interoperability of forces according to needs' (SMD, 2009).

13 Interview: Staff officer B, General Staff of the Army.

14 The description is based on SMD (2013). Words and sentences within quotes are literal translations by the authors of their equivalents in the Italian text. When available, we used NATO terms (NATO, 2008).

verification of facts and of the identified causes of the problem, and then of the proposed solution (if any). The second phase involves identification of 'further causes (implicit and second level) that produce the same effect'. The third phase consists in the elaboration of 'lines of action' through identification of 'remedial actions'. Once lessons have been identified, so that there is 'a theoretical solution' to the problem, an Action Body[15] and a Remedial Action (*azione correttiva*) are established to resolve the problem.

3. The *implementation* of lessons identified requires the Action Body to assign tasks (Action Plans) to various bodies (technically E/D/R, *Enti, distaccamenti e reparti*) of the Armed Forces and to monitor execution of the tasks according to the remedial actions proposed. The Action Body updates the leadership, for instance the *III Reparto*, Office for Lessons Learned of the Italian Army General Staff (Stato Maggiore Esercito, SME) through a Lessons Identified Report (LIDREP).
4. The *validation phase*, which moves from Lesson Identified to Lesson Learned, is undertaken by the Action Body with the support of the other organisations responsible for implementation. The validation is based on assessment of how Lessons Identified have performed in a variety of simulated scenarios. Further Remedial Actions may be suggested to address persisting problems.
5. The *dissemination phase*, in which Lessons Learned are then disseminated to the relevant bodies/units. The involvement of leadership is fundamental in this process so that the correct recipients of Lessons Learned are identified and transmission is smooth.[16]

Knowledge retention and transfer do not come about solely through the recently-formalized process described above. At the sub-national level, the importance of learning by individual brigades should not be underestimated. Each brigade has an officer for Lessons Learned – the army organizes specific courses for lessons learned staff officers – and this officer also contributes to sharing the documents produced at the end of missions; these documents take the form of case studies that can be disseminated across the service(s). Moreover, seminars are conducted by officers upon return from deployment, as well as informal meetings among members of the same units.[17] Joint training experiences have been perceived as very important in improving both interoperability and information sharing that can lead to the adoption of best practices developed elsewhere. In this sector, the

---

15 An Action Body is 'the organization or staff tasked with the implementation of assigned remedial action in association with a lesson identified. The action body develops an action plan to guide the remedial action activities' (NATO-JALLC 2011).

16 Interviews: Staff officer A, General Staff of the Army, Officer A, Italian Air Force, Institute of Military Aeronautical Sciences, General Commander, Taurinense Alpine Brigade.

17 Interview: General Commander, Folgore Parachute Brigade.

Army – inevitably given its disproportionate operational commitment – has been in charge of exercises.

At the supra-national level, namely NATO, the process is longer and more complex, but it is based on very similar premises (which in fact shape the national system). The institution of the Joint Analysis and Lessons Learned Centre (JALLC) in Lisbon in 2002 was a key step in standardizing the approach to Lessons Learned and in providing countries with a resource to develop their own national structures. JALLC in fact hosts the NATO database for Lessons Learned (LLDb), and it trains national officers who will occupy positions of importance in identifying and validating lessons at the national level. Italian officers have been trained by JALLC since the latter was created in 2010.

How Institutional Learning Happens

Endogenous and exogenous elements concur in bringing about change. According to the now classic thesis of Rosen (1991), the configuration of specific career patterns plays a role in shaping innovation. One of the factors that has changed the Italian armed forces in recent years is the advancement to higher ranks of officers with direct operational experience. As one officer interviewed observed, of central importance in promoting change has been 'the field experience of those who command, since this is the new generation that returns from operations and assumes commanding positions'.[18] Internal documents like the directive on careers issued by the SME recognize that experience abroad is both a key element in promotion and a preferred destination for the most highly deserving officers because it is an occasion for 'advanced education' and an 'investment' for the service (SME, 2011; SMA, 2012). Although career advancements still largely depend on the ranking obtained in military academies (which creates an element of endogeneity), previously non-existent or limited field experience is a factor important in shaping career patterns. The (virtuous) link among operational experience, promotability and credibility based on reputation acquired in operations has thus finally been established in the Italian Armed Forces (Rosen, 1991).

There is, of course, no necessary link between field experience and push to innovate (Adamsky, 2010), but it is likely that actual knowledge of field realities is the basis for better understanding of what solutions might effectively help, and thus a pre-requisite to promote innovation. On the other hand, the risks entailed by excessive reliance on officers solely on the basis of their field experience at the expense of other elements of their professional experience is not yet a major problem. Of course it may become one provided that the current standards of education (both at military academies and in postgraduate education currently required) are not maintained, as the decrease of resources devoted to training and education (mentioned in Chapter 3) seems to imply. Whilst the evolution of military education has been severely constrained by increasingly limited resources, it has

18 Interview: Officer A, Italian Air Force, Institute of Military Aeronautical Sciences

simultaneously become more important for career advancements and oriented to joint staff duties.[19]

Two limitations persist. The first is broadly linked to the relative openness of the armed forces to expertise deriving from civil society. Important steps have been taken to reform the 'reserve', which received major impetus in the 2000s (especially with legislative decree 66 of March 2010). This is coherent with new mission requirements that have emerged since the 1990s and with the focus on the comprehensive approach (with the relative need for civilian experts and liaisons) but it is still too early to evaluate the impact of the actual implementation of these reforms. On the other hand, the pooling of external resources has mostly involved personnel with non-core functions. Limited civilian involvement through joint military-civilian research and/or education centres and civilian participation in military think tanks still differentiates Italy from France and the UK (Rynning, 2003).

Internal resistances to change have occurred as well. Here, the major problem is the link between limited resources and force structure. As noted in Chapter 3, the number of effectively operational personnel in the Italian armed forces is limited. This creates a de facto obstacle to the effective flow of information on lessons learned and innovation. A two-tier structure has emerged: one tier consists of units with direct operational experience that interact – also through the newly-created mechanisms – with other units with similar experiences in order to share lessons and modify practices; the other tier is constituted by those units (mostly administrative branches) that have not benefited from field experience, and thus were excluded from the extra-ordinary budget resources devoted to missions that allowed further training and means availability.[20]

The external pushes for change have proved to be essential. The most direct effect of continuous deployment has been the possibility to observe and interiorize lessons and practices of others, principally the USA. The large asymmetry of resources, as well as of means employed, has inevitably led US armed forces to accumulate experiences and transform more rapidly. In Afghanistan in particular, the introduction of major changes in approach began in 2009 under the guidance of US General Stanley McChrystal according to the experience matured in Iraq and the new 'Petraeus doctrine'. As two officers interviewed (separately) noted,

---

19 Since the late 1990s, most officers assuming commanding positions at the joint staff level and within their services have attended the *Corso Superiore di Stato Maggiore Interforze* (Advanced Course of the Joint Staff) at the *Istituto Superiore di Stato Maggiore Interforze* in Rome, which includes officers at the rank of Major and Lt. Colonel of *Esercito, Marina, Aeronautica, Arma dei Carabinieri* e *Corpo della Guardia di Finanza*.

20 Interviews: General Commander, Pinerolo Mechanized Brigade, Officer A, Italian Air Force, Institute of Military Aeronautical Sciences, Officer C, Italian Air Force, Institute of Military Aeronautical Sciences.

Italian doctrine has 'trailed' NATO doctrine, and NATO doctrine has been largely shaped by ISAF experience in Afghanistan.[21]

The interaction between endogenous and exogenous factors enables more fine-grained analysis. Considering the elements mentioned in Chapter 2, the 'infectiousness' of US-originated doctrine in Afghanistan was high because of the perceived status of the organization that adopted them and the successes achieved in the early years of the adoption of new COIN techniques in Iraq after 2007. While the Petraeus approach to COIN was later subject to an increasing amount of criticism, to be borne in mind is the prestige of the American general, the Petraeus-inspired US Army and US Marines Field Manual 3–24 released in 2006 achieved in that period (Kilcullen, 2009).

'Proximity' also plays a key role. The existence of ties among organizations facilitated the exchange of information and the adoption of practices perceived by agents to be similar and legitimate. The overlapping of two patterns of interaction decisively contributed to learning in this context. As an officer observed, in fact, the Italian units in the field not only benefited from connections with and instructions from regional commands (at the tactical level), ISAF joint command (IJC, at the operational level) and ISAF HQ (at the strategic level), but they could also exploit the NATO structures in Europe,[22] with which they had a constant relationship outside operations. NATO made learning possible in various ways. On the one hand, directives issued by the Joint Forces Command in Brunssum updated SOPs on field realities, providing a source of reference for national forces. On the other, newly constituted structures within NATO provided countries with various arenas in which to promote and absorb innovation. The Allied Command Transformation, for instance, was of central importance in devising counter-IED practices, and the presence of Italian officers there greatly facilitated the process.[23] Intra-alliance sharing mechanisms allowed the training of officers in other countries where specific skills (both functional and geographical) had been developed,[24] and courses organized by the NATO rapid deployable forces at Solbiate Olona provided mission training that incorporated lessons from the experiences of other NATO countries as well as opportunities for informal discussion. More broadly, joint exercises within NATO have been found by the commanding officer of a brigade that has been traditionally very active to be a 'real engine of transformation'.[25]

In terms of 'susceptibility' to learning, motivation was extremely high due to the relative homogeneity of threats and the relative lack of operational capabilities of Italian armed forces. Together with motivation, in Afghanistan (and to a lesser

21 Interviews: Officer C, Italian Air Force, Institute of Military Aeronautical Sciences, Staff officer, NATO.

22 Interview: Officer A, Italian Air Force, Institute of Military Aeronautical Sciences.

23 Interview: Officer A, Taurinense Alpine Brigade.

24 Interview: Officer A, Italian Air Force, Institute of Military Aeronautical Sciences.

25 Interview: General Commander, Folgore Parachute Brigade.

extent in Iraq), the overall amount of resources was, by the standards of Italian deployment at least, considerable, which allowed the Italian armed forces to be receptive to innovations emerging in the theatre – also because Rome proved very responsive to requests from the field.[26] The other element is the units-core relation. Here the assessment of change is complex: as noted above, both the units (deployed) and the core changed, but the problems persisted in the transfer of knowledge within a service and among services not directly used in the field. This relates back to resources, which cannot guarantee the spread of innovations across the majority of forces.

As for the absorption capacity of the 'receiving' organization, prior knowledge has favoured the adoption of practices coherent with the 'Italian way of warfare'. As an Army officer put it, 'our approach was already attentive to the population', and therefore coherent with the population-centric approach that constituted an essential part of the new US COIN.[27] As noted in Chapter 4, the Petraeus doctrine spread by McChrystal through ISAF made that approach more systematic, but ISAF directives built on fertile ground within the Italian military, the majority of whom perceived it as a more organized framework for traditional Italian skills and ways of action (Coticchia, 2013). The Nibbio operation, established in 2003, was a smaller version in terms of set-up of duties and force structure of later US missions.[28]

Changing Force Structure: Assets

According to the Italian doctrine, future conflict will occur in complex operational environments characterized by 'hybrid threats' requiring adequate approaches and tools (SMD 2012). The crucial measures identified as possible guidance for 'force planning' are these: application of the 'comprehensive approach', the strengthening of Special Forces, the development of mobile and modular forces, and the provision of C2 facilities. Moreover, specific attention is paid to the 'dynamic and flexible procurement system' (SMD, 2012).

Regarding procurement, the new legislation adopted in December 2012 (Law 244) introduced several innovations, also with the aim of improving the (marginal) role of the Parliament. Indeed, the MPS of the Defence Commission had been involved almost solely at the very beginning of acquisition programme, whose costs frequently soared over time. Thus, the purpose of the new legislation was to strengthen the system of legislative control on the acquisition of armaments, fostering an enhanced flow of information (from the Ministry to the Parliament)

26 Interviews: General Commander, Folgore Parachute Brigade, General Commander, Pinerolo Mechanized Brigade, General Commander, Taurinense Alpine Brigade.

27 Interview: Brigade, Officer B, Folgore Parachute Brigade.

28 Interview General Commander, Folgore Parachute Brigade.

based on operational needs, strategic direction, development guidelines, and running programmes (Briani, 2012).

As illustrated by Chapter 4, the need to provide the proper equipment in risky military operations has been largely acknowledged by decision-makers. In addition, Italian defence has often modified the force structure deployed abroad in order to deal with new tasks or challenges (Afghanistan) or in the case of mission creep (Iraq).

Overall, mainly due to strategic uncertainty, European forces have maintained capabilities both for conventional warfare and low-intensity operations, which have represented the vast majority of the Italian missions undertaken in the post-Cold War era. Furthermore, as illustrated by Chapter 3, despite massive doctrinal attention to PSOs and constant involvement in those stabilization missions, 'conventional' long-term programmes have a primary function in the Italian defence budget, especially for the Air Force and Navy. Nonetheless, also the Army has had the opportunity to prosecute significant programmes related to equipment that have been crucial in low-intensity operations such as Afghanistan, Lebanon and Iraq: for instance, Lince light multi-purpose tactical vehicles, Spike third-generation antitank missiles, the SAMP/T antiaircraft missiles, counterfire radars, and protected remotely-controlled motorized turrets (Army Report, 2012).

As summarized by Farrell, Rynning and Terriff (2013), the literature on weapons acquisition illustrates how both bureaucratic interests and culture are important determinants of procurement. The authors stress the interests of organizations behind military programmes (such as the Future Rapid Effect System-FRES in the UK), but they also emphasize the importance of institutionalized ideas that have dominated procurement and doctrine development: for instance, jointness, the 'force for good' for the UK, and the preeminent French role in European security. What about Italy? On considering the interplay among doctrine, military programmes, and key assets adopted in operations, four features should be emphasized.

First, digitalization has been at the core of the reconfiguration and modernization of Italian forces. Above all, the NEC approach, and the consequent NEC Force programme, has shaped the Italian military with the aim of: 'achieving Information Superiority on operations and therefore a much greater decision-making capacity and higher performance than those of an opponent without similar resources' (Army Report 2012, p. 37). On a digitalized battlefield, all assets should be connected to foster synergy among all the capabilities, as suggested also by the 'Effects Based Operations' approach (Nones and Marrone, 2011). The key elements of the NEC doctrine are information sharing and situational awareness. Therefore, the goal of the NEC programme is the overall networking of the Italian armed forces, providing them with digitalized equipment such as *Soldato Futuro* (Future Soldier), SICCONA or SIACCON.[29] As already highlighted, the last two were

29 SIACCON is the *Sistema Automatizzato di Comando e Controllo* (Automated Information System for Command and Control).

crucial components of the VBLM Freccia deployed in Afghanistan. Several other expensive investments have been made for the purpose of broad digitalization, which is also conceived as a precondition for enhanced interoperability with allies – as evidenced, for instance, by the existence of joint programmes such as the SICRAL satellite. Moreover, analysis of operations on the ground has shown the central importance of UAVs, precisely to meet the requisite of better situation awareness. At the same time, the integration of command and control should also be enhanced among the three services (e.g. the Navy and Air Force adopt different C2s) and ISTAR capabilities are still required, as testified by the war in Libya.

However, in regard to digitalization, the best way to illustrate the evolution of procurement, operational requirements, and doctrine is to consider counter-IED warfare and the assets related to the protection of forces. The Italian forces arrived in Iraq and Afghanistan with VM 90s and Pumas, while these vehicles have now been replaced by VBLM Freccia and the VTLM Lince, whose production has been significantly increased after the positive results obtained on the ground against IEDs. A programme of modernization has shaped the Army in particular, enhancing existing tools (e.g., armoured vehicles) or providing new equipment such as the Guardian jamming device that provides electronic armour against road bombs (Ciampini, 2010).

In sum, the relation between digitalization and force protection highlights the interplay between interests (Italy has started to sell the Iveco Lince in several countries) and ideas (public opinion is deeply concerned about the safety of troops) behind weapons acquisition.

The second aspect that emerges on looking at the interaction among deployed military assets, procurement, and doctrine is 'power projection'. Italian defence focuses on the equipment that allows readiness and operational mobility (Graziano, 2013). Two examples help clarify this point. First, as stressed above, the Army has been structured as a medium-weight expeditionary force based on medium-weight vehicles providing constant mobility on the ground. Second, Italy has developed increased amphibious capabilities for its power projection, and the cornerstone of those capabilities is the new aircraft carrier *Cavour*.[30] Examination of the (very long) building process shows that the modifications which occurred in the design and planning were due to various factors: financial constraints and mounting costs, the changed 'operational and conceptual scenario' (Cosentino, 2010, p. 46) in the Mediterranean, Arabian Sea and Indian Ocean, and the modified structure of the Italian amphibious force, which was later considered to be adequate (as clearly demonstrated by the Operation Leonte in Lebanon) because of the creation of new units like the San Giorgio.

The third aspect of the relation between procurement and existing military programmes is the process of rationalization that has profoundly affected Italian defence in recent years. Political and military actors have raised alarm about

30 The first mission undertaken by the *Cavour*, after almost two decades of planning and ten years of building, was White Crane in Haiti (2010).

the effects of financial cuts on Italian defence. The former Minister, Mauro, warned of a 'functional default of the armed forces' (quoted in ANSA, 2013); the Undersecretary of Defence, Alfano, painted a bleak picture of force structure and equipment; and the Chief of Staff of the Navy, Admiral Giuseppe De Giorgi, even feared the extinction of the fleet, given the obsolescence of several units (Coticchia, 2014b). On the one hand, these concerns testify to the constant need to modernize armed forces shaped for decades according to the requirements imposed by the Cold War. Thus, the process of transformation has taken time, especially to change the 'crystallized' bipolar mentality.[31] Furthermore, numerous items of equipment still crucial for missions abroad (e.g. the AMX) have obviously been conceived for other strategic scenarios.[32] On the other hand, as already stressed in Chapter 3, equipment/investment expenditures have not suffered as much as operational and maintenance expenditures. Thanks to funds provided by the Ministry for Economic Development, several important military programmes have been developed in accordance with the initial plan. The ongoing debate on the F-35 (and FREEM and other large-scale, long-term programmes) has questioned specifically the costs of assets that have been so far guaranteed despite general financial constraints on the budget (Coticchia and Moro, 2014).

The last factor that has played a fundamental role in procurement, doctrine, and deployment is interoperability. We have already illustrated the importance of multinational interoperability in the transformation of Italian armed forces. Also in regard to the acquisition of new military assets, as well the overall shape of the force structure, the multinational dimension is of crucial importance, especially because of military cooperation.[33] The USA's influence has been constantly significant for Italy, as it has been for France and the UK (Farrell, Rynning and Terriff, 2013). However, like those countries, Italy's emulation has been partial. The literature shows how the extent of emulation depends on the financial, organizational and cultural capacities of states to absorb innovation (Horowitz, 2010). Farrell, Rynning and Terriff (2013) illustrate how financial constraints and diverse strategic views have affected the emulation process. For instance, the EBO became the effects-based operations approach (EBAO) in the UK, and the USA's net-centric warfare became the more human-centric approach to NEC adopted by the British armed forces. The same happened in Italy. As stated by the Army Report:

31 Interview: Officer A, Italian Air Force, Institute of Military Aeronautical Sciences.

32 Interview: Officer A, Italian Air Force, Institute of Military Aeronautical Sciences.

33 According to Goldman and Ross (2003), demonstration effects, military cooperation and professional education are the main mechanisms that give rise to emulation. Farrell, Rynning and Terriff (2013) believe that only the second is relevant to the cases of France and the UK. For Italy, all three mechanisms are significant because of Italy's different historical background and the absence of an autonomous Italian defence policy during the Cold War. This explains why the 'demonstration effect' and, above all, 'joint education' are crucial.

> The NEC approach differs from Network Centric Warfare (NCW). Unlike the latter, where all the assets employed have undergone a full digitisation process, the former envisages a step-by-step path to digitisation, as the assets are networked only when the technology on which they are based is sufficiently mature and reliable. Pending the maturation of the respective technologies, however, assets can also play a role in 'analogue' mode. (2012, p.37)

Obviously, Washington establishes the trend in regard to military technology, but from the Italian perspective the engine of the transformation is located within the NATO framework.[34] It is therefore important to look closely at the examples provided by allies such as France and the UK.

Learning and Innovation in the UK and France

Doctrine in Action

Chapter 3 illustrated the main elements of the British and French doctrines, while Chapter 4 examined the coherence between those aspects and military deployment. As summarized by Farrell (2008), the main innovations of post-2000 British defence pertain to jointness, the networking of forces, the effects-based approach to operations (EBAO), and the modular and expeditionary medium-weight nature of the forces. France shares all these features as the transformation process has been affected by similar external factors, such as the strategic change that occurred at the end of the Cold War and socio-technological innovation related to information technology (Farrell, Rynning and Terriff, 2013).

Analysis of the interaction between doctrine and deployment (from the MoD to the field and vice-versa) clarifies the crucial dimensions of the overall *process* of transformation and the *direction* of change. In regard to the coherence between official documents and operations on the ground, we can highlight two main features that have characterized the constant process of adjustment of force structure and strategic view. After focusing on operational experience, we shall examine the institutional dimension, also taking the procurement process and financial issues into account.

The first aspect is the role played by the comprehensive approach, which has become a 'guiding principle in the official military strategies and doctrines of many NATO countries, including Canada, France, Germany, Italy, the United Kingdom and the United States' (Alcaro and Marrone 2012, pp. 77–8). The ways in which the UK and France have shaped their approach to military operations have been based on a wider and holistic view of security. Within this broader framework, we can relate key concepts such as counterinsurgency or effects-based operations that

34 Interview: General Commander, Folgore Parachute Brigade. According to all our interviewees, the EU framework plays a very marginal, if not absent, role.

require a multidimensional focus on non-material aspects and increasing civil-military cooperation.[35]

In the case of the UK, the lessons learned from operations reveal the problematic adoption of those concepts, but also a capacity to adapt, modify, develop and integrate the doctrinal perspective. Iraq in particular showed the British difficulties in adopting an adequate COIN strategy. What has been termed the 'awakening of counterinsurgency thinking and doctrine-writing' (Egnell 2011, p. 299) testifies to the consequent attempt to revise and update approaches to guerrilla warfare. As stated by Ucko: 'Two broad lessons stand out: the British military's prowess with counterinsurgency is neither innate nor sufficient in the absence of a well-conceived and resourced strategy; and the British government's failure to prepare for and adapt to the challenges encountered in Basra reveal a low understanding of and priority for state-building' (2010, p. 132). According to Iron, the study of counterinsurgency had been ignored in favour of war-fighting in the bipolar era (at least until 2007), 'on the compelling, if incomplete, argument that an army prepared for high-intensity operations could 'step down' to other operations' (2013, p. 55). Hence the British approach was refined after the first years of military operations in Iraq and Afghanistan mainly due to operational experience and, above all, the US Petraeus manual. But whilst in the Middle East the British forces had already decided to limit their commitment, gradually withdrawing troops, in Afghanistan the UK started to deploy additional forces, embracing new concepts inspired by a renewed COIN strategy. The British campaign in Helmand therefore evolved 'from one centred on hard military power and directed at destroying the Taliban, to one focused on generating soft effects and securing the civilian population' (Farrell and Gordon, 2009, p. 673). The British approach changed in the different emphasis placed on non-kinetic effects and civil-military relations. New tactics and capabilities were developed. In sum, by adopting a more population-centric approach, the UK forces were able to adapt: they corrected their COIN strategy, their troops received more resources, and they were able to learn from past mistakes, although some problems persisted, both in terms of inter-force rivalry exacerbated by financial constraints and the operation's lack of political success.

As illustrated by Chapter 4, France, which for years had played a limited role in Afghanistan, started to increase its military commitment as a result of Sarkozy's Atlanticist foreign policy. According to the new geopolitical view established by the White Paper (2008), France intended to address the threats posed by regional crises from North Africa to the Middle East and beyond.[36] Thus readiness and mobility were vital for supporting a more active role in regional crises. Also in

35 All these terms are strictly related because of their conceptual similarity. For instance, Farrell, Rynning and Terriff (2013) consider the comprehensive approach to be part of the EBAO.

36 The new bases established (such as the new compound in Abu Dhabi) confirm this intention.

Afghanistan, France modified its policy in conformity with the new political and strategic perspective.

While Paris had for years refused to follow the paths of the allies (e.g. through PRTs), a different combat stance, new resources, and a closer focus on civil-military cooperation shaped the approach on the ground after 2008–2009. French forces tried to develop a 'national way to effects-based operations'. As the UK gradually developed EBO after Iraq (institutionalizing the EBAO concept), the French forces embraced the paradigm of the 'synergy of effects' (Alcaro and Marrone, 2012). In sum, a broader perspective to security context was the guideline for the armed forces, which were not employed for the mere destruction of the enemy. However, French efforts in Afghanistan remained limited (despite a dramatic cost in terms of casualties), especially because France was one the first ISAF contributors to plan its withdrawal, thus reducing the time scale on which it is possible to assess the new approach in the field.

The second general feature of the interplay between doctrine and deployment concerns the efforts made by British and French forces to modernize their equipment and move towards the overall digitalization of military units. As in the case of Italy, NEC programme were gradually institutionalized in the British armed forces through a complex process that encountered several obstacles. Apart from the financial constraints of the 'age of austerity' (Cornish and Dorman, 2009), operational demands severely affected the concrete implementation of what the official documents envisaged in terms of net-centric capabilities: indeed, the protection of troops required new equipment, such as armoured vehicles and helicopters.[37] For instance, the scant progress of the FRES armoured fighting vehicles programme was due to the priority given to the urgent needs emerging on the ground. Also in the case of France, the planned digitalized force faced difficulties when it was required for rapid action – as demonstrated by the first months of the operation in Afghanistan.

Notwithstanding the problems encountered by both forces throughout the process of modernization, they gradually institutionalized the digitalized units, widely adopting the expeditionary nature designated by official documents. Adaptability and flexibility became crucial for handling manifold conflict scenarios, as illustrated by the different natures of the operations undertaken in Iraq, Afghanistan, Lebanon and Libya. As stated by Cornish and Dorman: 'transformation should be a means to an end, a journey from one set of capabilities to another' (2009, p. 743). Unforeseen changes in the strategic environment, technological evolution, and budget cuts created both constraints on and opportunities for transformation, and they shaped the direction of innovation. Overall, British and French forces gradually followed the main path established by doctrine. But the demands from the ground were crucial for the adaptation

37 Proper armoured vehicles (mainly the VBMR – *véhicule blindé multiroles* and EBRC – *engine blindé à roués de contact*) acquired centrality also for the French forces, as in the cases of Italy and the United Kingdom.

of forces, as well as for refining the concepts drafted in official documents. The needs and lessons learned from the field were also fundamental in the broader institutionalization of change.

Institutional Change

The UK and France also experienced a process of 'transformation in contact' similar to the Italian one (the expression has been used by British General David Richards, quoted in Foley, Griffin and McCartney, 2011). Even in these cases, the last decade or so was decisive in the setup of more institutionalized systems for lessons learned (or RETEX, *Retour d'experience* in France), and even here the Land Forces were at the forefront of the process.

In the UK, the process of building a systematic approach to doctrine and training incorporating lessons learned had started in the late Eighties.[38] In 1994, the Directorate General of Doctrine and Development (DGD&D) was formed with the responsibility for doctrine, experimentation and force development in the British army. In 2006, the DGD&D was substituted by the Development, Concepts and Doctrine Centre (DCDC), a structure at the joint level reporting directly to the Chief of the General Staff, signalling a strong push towards jointness, but a formalized structure to deal with lessons learned was still lacking. In the wake of the invasion of Iraq in 2003, General Sir Timothy Granville-Chapman highlighted the lack of a formal process within the British army to collect and analyse lessons from operations. To correct this deficiency he issued the first *Land Standing Order 1118: the lessons process*. Responsibility for lessons management was placed firmly in the hands of a two-star general, Chief of Staff (COS) Land Forces, and the Mission Support Group (MSG) was strengthened. Central to the MSG's work was engagement with every British brigade and supporting unit going to Iraq. The team conducted thorough analysis of all brigade reports, especially the crucial post-operation reports. Also, the new order formalized the procedures for post-operation interviews with brigade commanders and principal staff officers returning from operations. In 2006 the MSG was folded into the new Warfare Development Group.

The development of RETEX in France dates back to the Nineties, after the intervention in the Gulf War (Durand, 2001; Hotier, 2002; Brustlein, 2011), but the current structure took shape in the following decade. In 2003, a joint centre for RETEX had been created at the État Major des Armées, while the Army had already in the previous years dedicated structures to that task. In 2000, a *Centre d'évaluation et de retour d'expérience* (CEREX, Centre for Evaluation and Lessons Learned) had been constituted, and in 2004 the newly-created *Centre de Doctrine de l'Emploi des Forces* (CDEF, Centre for Doctrine and Force Employment) took the responsibilities for RETEX, creating an internal division tasked with that duty

38 A more complete account of institutional transformation of the British armed forces can be found in Foley, Griffin and McCartney (2011).

(*Division Recherche et Retour d'Experience*, DREX). The Air Force and the Navy came later in the process, but devoted lesser resources. The Air Force created a centre for RETEX within the État Major des Armée de l'Air (EMAA) in 2008, and the Navy a bureau within the État Major des Opérations (EMO) de *l'État Major de la Marine* in 2010 (Brustlein, 2011).

Also in the cases of the UK and France, jointness has been a crucial feature of the transformation process. As already stated, new joint institutions have been created, such as the Permanent Joint Headquarters (PJHQ) and the Joint Doctrine and Concepts Centre (JDCC), which have assumed greater control over British doctrine. In France, the Joint Improvised Explosives Device Defeat Organization (JIEDDO) was conceived as the institutional answer to the growing inter-force needs regarding the threats posed by IEDs (Perrin, 2009).

In the UK, after publication of the Levine Report (2011) the three military service chiefs were removed from the Defence Board (where decisions on equipment are taken), thereby increasing the role of the Chief of the Defence Staff. According to Farrell, Rynning and Terriff (2013), joint institutions have significantly shaped the entire process of transformation, also because of the role played by the new Joint Force Command. For Foley, Griffin and McCartney (2011), the joint defence organization, which was also viewed as a way to rationalize and save costs, has reduced the autonomy of individual services in terms of planning, training and education functions, while services have been left with low-level responsibilities.[39]

Besides changes in the macro-institutional setup, several elements emerging from operational experience have been challenging other existing institutions and practices. With reference to inter-operability, the UK places great attention in working with the US, with constant exchanges that go far beyond the NATO standards. The level of information sharing is constantly high, as is working together in complex terrains, such as the Helmand province in Afghanistan, and in performing operations using US platforms (including flying from US carriers), though such cooperation means that there does not entail full parity, real or perceived (Marston, 2008; Taylor, 2009).

Interoperability with allies has grown increasingly important for France as well. Under the Sarkozy presidency there was a push for 'rapprochement' with NATO and the USA. Several examples can be cited to illustrate this change. France deployed troops in new multinational bases (e.g. Djibouti, together with US Marines) in strategic regions, while it committed its Afghan regional command under the United States. France enhanced its partnership with the Atlantic Alliance in terms of shared communication systems, but several problems remained, such

39 According to the authors, this process had the consequences of reducing the British army's ability to learn lessons from training and operations. 'The Army's "brainpower" was increasingly absorbed with the challenges of "higher-level" doctrinal and conceptual development and with providing middle- and senior- ranking army officers with appropriate education within joint establishments' (2011, p. 261).

as the impossibility of exchanging data directly with NATO allied aircraft because French fighters do not use Datalink-16 (Alcaro and Marrone, 2012).

Apart from the NATO framework,[40] one of the most potentially far-reaching events in terms of interoperability has been the agreement signed by France and the UK in 2010. This should lead to significant change in their deployable headquarters capability by setting-up a Combined Joint Force Headquarters in order to operate the previously envisaged Combined Joint Expeditionary Force. In the future, as the Combined Joint Expeditionary Force should get to full operational capability around 2016, shifting the emphasis of Franco-British co-operation in theatre from de-confliction to greater interoperability will be the revealing test by which people come to realize the extent to which there is growing substance behind the treaties.

Overall, Britain, France (and Italy) have increased their convergence on the US model: net-centric capabilities, expeditionary power projection, modularity, and centralization of command (Dyson, 2008). Moreover, the creation of Joint Reaction Forces in London and Paris has mirrored the USA's integration of services (which is obviously crucial for participation in multinational coalitions). As stated in previous sections, the emulation has been partial (from NCW to NEC, from EBO to EBAO or synergic) but inter-organizational learning has still been significant. Infectiousness, proximity, susceptibility, and absorption capacity are all important mechanisms in this regard. Indeed, working with US allies in a multinational framework like the ISAF has exposed European officers to best practices and learning. Account should also be taken of the claimed (though questionable) 'success' of the US COIN in Iraq, the standard common procedures adopted during joint missions on the ground, the training promoted at NATO level on crucial concepts first developed in Washington (from digitalization of the battlefield to effects-based operations), the development of a shared mentality after decades of multinational interventions, and the strong motivation to learn in order to be important partners (military, political and economic).

Current institutions have been considerably affected also by the need to refine existing concepts or to develop new strategic approaches through transformation of the main training programmes. Chapter 4 described how the British Army encountered significant problems in the practical implementation of key counterinsurgency concepts in Iraq and Afghanistan. According to Marston (2009), the army 'lost its way' because of the lack of education in COIN from RMAS through to Staff College. For Little (2009), the UK forces (wrongly) believed that they had more to teach than to learn regarding COIN. Farrell, Rynning and Terriff (2013) confirmed the slow British institutionalization

40 Dyson (2008) rejects the pessimistic picture regarding the lack of complementarity between the EU and NATO. He highlights the increasing cooperation in terms of military capacities and force planning, as illustrated by the Berlin-Plus Agreement (2003). Indeed, EU battle-groups are lighter and focused on low-medium intensity crisis-management mechanisms. The strategic coherence between NATO and the EU (sharing approaches and threats perceptions) has been demonstrated in several official documents in recent years.

concerning COIN (as testified by the delay in developing a brand-new COIN doctrine) and the absence of proper mechanisms to exploit operational experience in the early 2000s.

Operational pressures were crucial for the revision of army learning mechanisms, which were 'under-resourced' in comparison to the massive military commitment abroad (Foley, Griffin and McCartney, 2011). Therefore the purpose of creating the Lessons Exploitation Centre (LXC) and the Army's Land Warfare Centre (LWC) was to fill this gap. By recombining doctrine, development and training, 'the British army took a conscious decision to make the structural changes necessary to enable wider army transformation' (2011, p. 262).

In the case of France, several authors emphasize the growing attention paid to training and validation before deployment (Givre, 2009; Perrin, 2009). The support provided by units returning from missions (within the foreign military training unit – DIO), as well as the After Action Review (AAR), which can help to address mistakes and to disseminate lessons learned, have been factors crucial in enhancing the training process (Perrin, 2009). Other authors, while acknowledging such efforts,[41] stress the need to improve the role of simulation in training as a way better to prepare troops for high-risk engagements (Florin, 2009).

However, Haddad (2010), who has analysed the 'cultural training' of the French forces relative to their deployment in Lebanon, reports general satisfaction among troops with the type of preparation that they have received. But whilst intercultural and social aspects have been important and appreciated aspects of the curricula, French soldiers complain about a lack of specific training in the UN system, which is viewed as extremely complex (and this helps explain why, as for the Italians, the NATO umbrella is largely preferred).

Improvements have been significant in specific sectors as a consequence of operational demands. In this sense, the military operation in Afghanistan has been a game-changer.

Both in Britain and France the integration of counter-IED exercises has been 'systematic' (Perrin, 2009, p. 55). But this happened only some years after the beginning of military commitment in Central Asia. In late 2009, the Dixon Review of NEC and anti-IED capabilities still stressed shortcomings in pre-deployment training. Also in the case of France, mainly due to the different level of engagement, the integration of counter-IED exercises occurred after 2009.[42]

Foley, Griffin and McCartney stress the remarkable capacity of the UK forces to review, revise and update training (and doctrine) in order to 'counter changes

41 As reported by Perrin: 'The Land Forces Command carries out studies about the details of implementation to ensure the best training possible at all levels for all services' (2009, p. 55). The operational preparation for deployment (MCP) has assumed a vital role at the Forces Training Centre.

42 As stated by Marrone and Alcaro (2012), the Allied Command Transformation (ACT) supervised an action plan both to provide the systems and technologies needed to counter IEDs and to improve the relative training of troops.

in the enemy's own tactics (most notably, the continual revisions of improvised explosive device planning and sniper activity)' (2011, p. 262). The authors are more doubtful about translation of battlefield lessons into more widespread change in the army because of the lack of 'a coherent command structure that linked the lessons-learned system with doctrine and force development, the British army's lessons system' (2011, p. 262). Yet the main problems and limitations related to training stem from cuts made to the defence budget.

As already illustrated in the case of Italy, the overall process of institutional change has been profoundly affected by financial constraints in both France and UK. Two main features of the interplay among procurement, military programme and transformation are to be stressed.

The first is 'operational overstretch'. As illustrated by Chapter 3, France and (especially) the UK made massive cuts to personnel after the end of the bipolar era.[43] UK transformation occurred in the context of a growing mismatch between commitment and resources. While the budget was cut (in real terms) and the number of frontline units, particularly for the Army, were reduced, the UK's international commitment increased in two risky operations: Iraq and Afghanistan. Michael Clarke, director of the Royal United Service, described the British Army as an 'engine running without oil' (*The Economist*, 2009). According to Cornish and Dorman, the government struggled to reconcile 'current operational costs with projected expenditure on a range of very large defence acquisition projects, and to do so while a deep economic recession looms' (2009, p. 261). During a period of austerity, therefore, European countries were considerably involved in intense missions requiring new tasks, approaches, and tools. The transformation process was consequently influenced by the financial dimension, which obviously interacts with the interplay among operations, doctrine and institutions.

The inevitable inter-service rivalries well illustrate this dynamic relation. Chapter 3 has already highlighted the evolution of defence budgets in Italy, France and the UK. For instance, in Britain 'heavy spending on kit for the navy and air force leaves little for the army; one source says it will receive less than 10% of all spending on defence equipment between 2003 and 2018' (Cornish and Dorman, 2009, p. 261). On the one hand, procurement traditionally aims at maintaining a wide range of military capabilities. Thus the costs of large-scale programmes for the navy or the air force have been significant owing to the size and role of the armed forces. On the other hand, operations have driven change in the armed forces from below, diverting core defence expenditures to ongoing missions (Edmund, 2010). Intense military activity has had the consequence of increasing the focus on the short term, thus on keeping up with daily operational requirements, with the possible risk of 'a tyrannical effect on defence equipment acquisition' (Cornish and Dorman 2009, p. 738). Several examples help illustrate

43 According to UK Defence Statistics (2008) and DASA (2009), the UK military services suffered a 39% cut in the period 1990–2008. Reported in Cornish and Dorman (2009).

this process, for both France and (mainly) the UK. For instance, especially due to the above-mentioned Urgent Operational Requirement, the British government (which used the Treasury's reserve funds) delivered better-protected transport vehicles and other tools to guarantee adequate equipment for the troops (King, 2010). Indeed, as described by Farrell, Rynning and Terriff (2013), the Snatch Land Rovers provided no proper protection for soldiers at the beginning of the missions in Afghanistan, and then, thanks to UORs, Mastiffs were deployed. All NATO allies were forced to develop new capabilities such as jamming systems to counter IEDs, while ISAF evidenced the scant availability of fixed-wing and rotary-wing air capabilities.

At the same time, as already mentioned, programmes like FREEM lost their momentum principally because of those short-term investments.[44] Also France suffered important delays in large-scale programmes, such as the fourth strategic submarine, while the second carrier was postponed. Paris, mostly due to the new combat standing in eastern Afghanistan, focused strongly on expeditionary capabilities, satellite capabilities, and air-to-ground support.

Moreover, the operations were also opportunities to review former decisions on cuts. Thanks to the UK mission in Libya, the time in service of the Nimrod R1 reconnaissance aircraft fleet (which was due to be retired in March 2011) was extended. According to Cenciotti (2011), the air strikes showed that assets scheduled for retirement (e.g. the RAF Sentinel R1 spy-plane) were still efficient and relevant. Apart from Libya, all regiments wanted to demonstrate their skill because operational performance is crucial to institutional survival in a context of limited resources.

Overall, inter-service rivalry, which was clearly related to defence budget cuts, has increased attention to financial issues in Great Britain (as well as in France and Italy). More specifically the debate on defence has focused almost exclusively on inputs (like military equipment), rather than on outputs (the general aim of defence policy). But according Cornish and Dorman, the defence debate is about 'using armed forces to achieve value in the form of specified political purposes, and doing so with limited resources' (2009, pp. 752–3). The interplay among different dimensions (budget, doctrine, operations) has shed light on the *direction* of military transformation, highlighting political aims, interests, values, obstacles and opportunities provided by military operations.

44 By contrast, in the case of France, a similar expensive programme like 'Scorpion' had a different process implementation.

Chapter 6
Conclusions

Adapt, Improvise, Overcome?

The extent of the transformation of post-Cold War Italian defence policy has been massive. Indeed, Italy has profoundly modified institutions and doctrine, making its military contribution through the constant commitment of forces in multinational operations abroad. Dyson (2008) identifies a convergence between the aims of British and French defence policy after the end of the bipolar era as they shifted from the core objective of territorial defence towards expeditionary crisis-management missions to deal with multidimensional threats to international instability. Italy followed the same trajectory, finally adopting a more dynamic defence policy after years of inaction due to the bipolar (domestic and international) constraints. Thus, specifically because of this historical background, Italy's evolution has been much greater. The exceptional advance of logistics, which have been substantially altered and improved in order to support wider military activism abroad, is one of the best examples of this innovation.

However, several elements also point out to continuity, such as the enduring role of traditional values like 'multilateralism' (Ignazi, Giacomello and Coticchia, 2012). Another example illustrates the complex role of Cold War legacies in the case of Italy: the AMX fighter-bombers, which returned from Afghanistan in June 2014 after five years of operations (the longest mission for Italian aircraft since WWII). Here, two elements help elucidate the complexity of the process of military transformation in Italy. Firstly, despite the importance of the task undertaken by the AMXs, no information about targets, air strikes and operations has been issued. In other words, widespread indifference towards military affairs still affects the national debate. Secondly, the Italian Air Force employed an aircraft planned and developed decades ago in a completely diverse strategic environment. In general, Italy has constantly adopted (and accordingly adapted) equipment previously acquired for different purposes.

However, something has changed in the past few years. On the one hand, the debate on recent reforms (such as the so-called 'Spending Review Act' – Law 244 of December 2012) has increased attention to the rationalization of the forces and structures of Italian defence. On the other hand, the controversy surrounding the F-35 JSF (or the FREEM frigates) has revealed the broader programme of equipment renovation strongly advocated by the armed forces (especially the Navy and the Air Force).

Nevertheless, both of these apparent novelties are strictly related to the financial dimension. In other words, are these discussions, as well as enactment of the reform, budget-driven? The rationalization of spending is certainly crucial in a country hard hit by the financial crisis. As highlighted by the Army Report: 'The review process will no longer rely on capability-driven planning, but rather on a more pragmatic approach to continuous assessment of the economic sustainability of the solutions adopted' (2012, p.114). General Graziano (2013) confirms that the economic sustainability of the armed forces is now the core principle for planning. But, as already stated in Chapter 5, the risks of a budget-driven approach are manifold, especially as regards possible underestimation of long-term strategic and political prospects. Also for this reason, although financial considerations are still crucial constraints in Italy, the current debate on a (finally) new White Paper seems to give higher priority to the design of broader strategic guidelines so that the entire review can be properly conducted.[1]

Crucial for compiling the new White Paper is accurate assessment of the complex process of transforming the Italian armed forces in light of the main lessons learned in decades of military commitment. But the public debate as well as the national literature on this issue is still extremely scarce. This book has sought to fill this gap by showing the main features of the transformation. The goal has not been to identify the deep causes of change, but rather to illustrate the transformation process by focusing on the dynamic interaction among its main dimensions: budget, doctrine and deployment. As suggested by Farrell, Rynning and Terriff (2013) external factors –such as post-bipolar strategic change or the socio-technological innovation brought about by information technology – have been vital in explaining transformation. But in order to understand the direction of change, we should look in detail at the complex interplay among diverse elements.

Although the armed forces of European countries have been extremely active in recent decades, analyses considering the full spectrum of the change process – from doctrine to budget and force structure, and then from operational experience to institutional learning – have been mostly restricted to the major powers. Other countries have often received less attention; and Italy in particular has been neglected by most research. Moreover, the existing studies tend to take change for granted and do not explore how the change process unfolds. The book has reconstructed how transformation has taken place in Italy and the extent to which it exhibits patterns similar to those of the UK and France.

Analysis of the Italian case has highlighted the crucial importance of deployment for the innovation process. Interaction with the other dimensions reveals the central importance of missions abroad in several respects. Firstly, operational experience has guaranteed a constant flow of economic resources in a period of austerity. Thanks to specific decrees issued to finance operations, the

1 See, for instance, 'Guidelines – White Paper for international security and defence', available at: http://www.difesa.it/EN/Documents/White%20Paper%20-%20Guidelines.pdf. The 'Guidelines' is a preliminary analysis conducted by the Group of Experts appointed by the Minister of Defence (June 2014) in order to collect comments, perspectives and data as the basis for the next phase of developing the White Paper.

Italian armed forces have been able to maintain their capabilities, also in terms of training. In fact, the scrutiny of budget trends conducted in Chapter 3 highlighted a paradox. Whilst in the past resources were relatively abundant in a situation of virtually non-existent operational commitments, the situation has now reversed: exactly at a time when the military is constantly employed, there is no parallel abundance of resources.

Decreasing resources might further exacerbate the inter-service rivalries for their allocation. The competition between the Air Force and Navy over the acquisition of the JSF platforms, for instance, is likely to be intensified if the number of planes is further reduced in the future. A key issue that emerged in recent years (in parallel with other NATO countries) is the disproportionate role played by the Army compared with other services, with consequent needs and demands that have acquired priority in the debate (e.g., armoured vehicles and other tools to enhance the safety of troops). Indeed, mirroring international experiences, attempts to promote change in the armed forces in light of Italy's long experience in Afghanistan – thus through institutionalization of lessons learned in regard to COIN – may generate conflict (while inter-service coordination problems persist).

Secondly, the role of deployment has obviously been essential for lessons learned, in terms of both emulation of allies on the ground and to shape the change of the institutional framework devoted to this issue. In fact, the Italian armed forces have begun a process of institutional learning that may contribute decisively to maintaining organizational memory and specific capabilities. Although it is probably too early properly to assess a learning system that has been implemented only in recent years, some aspects are already apparent. For instance, operational requirements have noticeably altered the training system and its features (e.g. the new centres focused on anti-IEDs). More broadly, the role of learning has become both more central and more difficult because the adoption of new standards frequently requires additional economic and conceptual resources. In this regard, if the influence exerted by the NATO framework is not considered, it is impossible to comprehend the direction of transformation in Italy, especially in terms of the systematization of learning and doctrinal review.

The third dimension is precisely doctrine. Despite limited production over time (no White Paper was issued for twelve years after 2002) several minor documents testify to how Italian defence has gradually included, modified, and updated key concepts proving crucial in the field, such as the comprehensive approach of net-centric capabilities.[2] The Italian army has been a 'pioneer in adopting a comprehensive approach that fostered development and stability' (Army Report 2012, p. 15). The concept of counterinsurgency well explains how the interaction among different conceptual and operational dimensions has fostered the transformation of forces. On the one hand, Italy had already implemented on

2 Italian military doctrine also focuses on a synonym for the comprehensive approach: the so-called DIME approach, which is based on four components of national power: diplomacy intelligence, military, economy.

the ground (since the 1992 operation in Somalia) key components of the current US-driven COIN approach, for instance by developing civil-military units and military-police forces. Indeed, national documents have constantly emphasized that these are the crucial components of the Italian approach to intervention. However, on the other hand, these aspects were not systematized into a wider COIN doctrine before troops were sent to Afghanistan and Iraq. It is for this reason, according to several officers interviewed, that most Italian forces considered the COIN approach developed by McChrystal within the ISAF framework to be their main referent in terms of counterinsurgency.

In sum, we should interpret the transformation as a complex configuration of factors where bottom-up military adaptation has been primarily affected by experience on the ground, and where the allies' influence has also been massive. This interplay is not limited to the field; it also occurs at a higher level owing to the role played by multinational frameworks, especially NATO, in shaping learning, doctrine and institutional structures. This process is modified and reinforced according to domestic perspectives and needs dominated by financial considerations (especially since the onset of the crisis). For Farrell, 'military innovation is a dynamic process precisely because it involves the interplay of international drivers and national shaping factors' (2008, p. 783). The entire process is extremely dynamic. Therefore, the above-mentioned elements are not alternatives in affecting change; rather, they are all vital factors in a multifaceted process of transformation.

Changing with Limited Resources: A Comparative Perspective

Consideration of the co-evolution of French and UK doctrines, defence management, power projection capabilities and financial restraints provides an additional perspective from which to assess and understand Italian military transformation. Indeed, procurement, institutional modification and strategic concepts exhibit similar patterns, although to different extents.

As Strachan writes: 'Since 2003 the length of the Iraq and Afghan campaigns, and the disproportionate demands these campaigns have placed on the army over the other two services, have created tensions within the armed forces' (2011, p.63). While land warfare has been at the centre of strategic reflection, increasing resources have been devoted to Urgent Operational Requirements in order to meet short-term needs on the ground. Tensions have been created by the gap between the long-term procurement[3] budget and the needs relative to those urgent requirements (mainly for the Army). On the one hand, it is possible that

3 Despite notable achievements in the past two decades, the recent failure of the attempt to create a more competitive market through privatization of the sector (because of the lack of interest among companies) reveals the need to improve the procurement programmes further.

the growing expenditure on Urgent Operational Requirements has caused delays and cost overruns in long-term programmes (a classic case of prevalence of the logic of exploitation over exploration). However, some authors have underlined that cutting conventional capabilities for the Army has been easier than reducing complex large-capital development programmes (e.g. destroyers, submarines, etc.) that for years 'tie the hands'[4] of the Navy and the Air Force. Domestic reasons therefore hinder cuts in both directions. Public opinion and political parties care about the cuts in personnel (which concern a specific community and territory and are naturally more severe for the Army because of its composition). Above all, they are concerned about large-scale equipment programmes involving numerous workers (and thus are difficult to reduce for electoral reasons). In fact, despite financial constraints, numerous 'big programmes' have been preserved. Finally, as a way to manage and (ideally) overcome inter-service rivalry, joint institutions have gradually increased responsibilities for developing key concepts and managing the lessons learned process.

Jointness, digitalization, modular and medium-weight expeditionary forces are key concerns also for France, where the political shift from Chirac and the *Model 2015* to Sarkozy and his new pro-US and Atlanticist policy has been a crucial turning-point in the innovation process. As testified by operational engagement, it has affected the direction of change. Finally, financial constraints have deeply influenced the extent of the innovation envisaged by the ambitious *Livre Blanc 2008*.

Overall, the limits set by the economic crisis on transformation have induced several analyses to focus primarily on the risks due to shrinking budgets. However, the forthcoming challenges for Italian and European armed forces should be conceived from a broader perspective. The book goes beyond the current problems posed by financial limitations, examining the co-evolution of three crucial dimensions of transformation: budget, deployment and doctrine. Despite mostly devoted to the Italian case, the comparative outlook of the volume provides crucial landmarks in order to better understand the whole process of innovation. Italy, France and the UK have guaranteed the most significant military contribution to Western missions abroad in the new century. A constant search for appropriate doctrines and equipment deeply marked their evolving defence policy. In fact, the three countries have steadily adapted their armed forces while being involved in complex operations that altered the general perception of contemporary warfare. The 'war amongst the people' became the challenging scenario for thousands of Italian, English and French troops (Smith 2006). A comparative perspective sheds more light on the interaction among the key-components of transformation, illustrating the ways through which significant changes have been contemporaneously faced in Rome, London and Paris.

4 Interview: Farrell.

Learning and Innovation: Challenges and Ways ahead for Analysis

All three of the words used in the title of this chapter ('adapt', 'improvise', 'overcome') co-exist in the process that we have analysed. For instance, Italian forces have constantly adapted structures, approaches and equipment in order to deal with emerging threats like IEDs.[5] Troops have often improvised – as they did during the first weeks in Afghanistan when Italian soldiers asked to change their dress uniforms because they were dangerously similar to those of the Soviets. While these two elements are now matters of fact, 'overcoming' is still a major question. Italian decision-makers have conceived 'Smart Defence' or 'Pooling&Sharing' as vital if Italy is to preserve and develop capabilities even in a period of austerity (Coticchia and Moro, 2014).

However, as stated above, there are challenges other than limited resources. If missions abroad are reduced (without re-allocation of funding towards training and validation of lessons learned), it may not be possible to maintain innovation – also because only the deployable component of the Italian armed forces has had operational experience, while the other sections (non-deployable) are still legacies from the Cold War. Moreover, the forces are still prepared for a wide range of missions across the whole spectrum of operations and conflicts, from peacekeeping to high-intensity warfare.

The Army Report (2012) states that Italian forces will maintain their joint and expeditionary nature according to the 'transform while operating principle' (2012, p.105). However, there is little doubt that the end of a major military commitment like Afghanistan will leave armed forces in Italy and elsewhere in need of thorough restructuring. From a wider viewpoint, there is still a relative lack of debate on military issues in Italy. This is all the more evident at the operational level, which often escapes broader analysis of errors committed and lessons learned.

Of course, we are conscious that the book – which aimed to provide an overview of force transformation of the Italian armed forces – falls short of providing definitive answers on important theoretical questions and policy challenges as well. We believe, however, that it paves the way for further research and analysis in both fields, and we want to conclude by indicating possible directions. From a theoretical perspective, further inquiry on the mechanisms of learning can proceed through more disaggregated analysis of how different agents in the field (different units, even belonging to different countries) have adapted in front of similar circumstances, so to single out which specific factors underpinned change or resistance. In particular, traditional studies on military innovation stemmed from the need of explaining change in the armed forces of great powers, while current structure of international politics also requires deeper understanding of the mechanisms that drive (and describe) transformation of 'second-rank powers' or 'middle powers' For instance, this type of research is coherent with

---

5 The Army Report identifies a 'culture of adaptiveness' that has shaped Italian forces (2012, p. 47).

the need for multinational contingents, starting from NATO and its partners, to co-operate and inter-operate, a recurring feature of military missions of the last two decades. Better grasp of this phenomenon of transformation requires then looking more thoroughly at sources of inter-organizational change (summarized in Chapter 5). Countries such as Italy, France and the UK (but the list here would be much longer) adopt doctrines, institutions, procedures that represent the result of a complex mediation among several factors. Specifically, while even for great powers both environmental pressures deriving from the operational environment and domestic and organizational features interact in shaping transformation, the armed forces of other countries do adapt and learn (or at least try to) by looking at what others (the US to begin with) do. This 'extra' source of change provides a useful tool to complement theories of foreign policy behaviour concentrating on middle-powers. Although more can be certainly done to disentangle the logics that drive processes of isomorphism (as well as allomorphism) among different armed forces, this book highlighted how themes important in contemporary as well as traditional IR theory can be addressed by looking at armed forces transformation from a 'bottom-up' perspective.

In terms of policy, several doctrinal and organizational choices and decisions on budget allocation can emerge, also given the requirements dictated by different policy preferences based on visions of future challenges. Drafting a new White Paper in Italy is then a process that deserves a fair degree of emphasis as the document will make such policy preferences on national doctrine more transparent and as such constitutes the premises for expanding the much needed public debate on the subject. This is all the more true in the current global security environment, where multidimensional and rapidly evolving threats create an ambiguous environment for armed forces. Nevertheless, the key lesson of the past conflicts is that the armed forces that begin the operations are different from the armed forces that emerge from them. Winning the next war starts from the recognition that it is essential to develop military doctrine that clearly states the ends and means of defence policy so that their balance is sustainable over time. But it does not end there. What should not be forgotten, however, when the new White Paper will be completed, is that transformation is a continuous process and that a single, even smart and comprehensive, reform cannot substitute for a process of continuous self-assessment.

References

Adamsky, D., 2010. *The Culture of Military Innovation. The Impact of Cultural Factors on the Revolution in Military Affairs in Russia, the US, and Israel.* Stanford: Stanford University Press.

Angioni, F., 1984. *Un soldato italiano in Libano.* Milano: Rizzoli.

ANSA, 2009. *Afghanistan: arrivano i 'Lince' rinforzati, ma i militari sono perplessi.* 7 October.

Arbuckle, J.V., 2006. *Military Forces in 21st Century Peace Operations.* New York: Routledge.

Argote, L. and Myron-Specter, E., 2011. Organizational learning: From experience to knowledge. *Management Science*, 22(5), pp. 1123–37.

Armacost, M.H., 1969. *The Politics of Weapons Innovation: The Thor-Jupiter Controversy.* New York: Columbia University Press.

Army Recognition, 2012. *French Armed Forces in Lebanon.* 27 January.

Arquilla, J. and Ronfeldt, D., eds., 1997. *In Athena's Camp. Preparing for Conflict in the Information Age.* Santa Monica, CA: Rand Corporation.

Avant, D., 1993. The institutional sources of military doctrine: Hegemons in peripheral wars. *International Studies Quarterly*, 37(4), pp. 409–30.

Battistelli, F., Galantino, M.G., Fay Lucianetti, L. e Striuli, L., 2012. *Opinioni sulla Guerra. L'Opinione Pubblica Italiana e Internazionale di Fronte all'Uso della Forza.* Milano: Franco Angeli.

Battisti, G., 2007. Le fortificazioni nella controguerriglia. *Rivista militare*, 4, pp. 48–61.

Battisti, G., 2012. Il Volo del Nibbio. *Rivista Militare*, 4, pp. 74–8.

Bellamy, A. J., 2009. Realizing the responsibility to protect. *International Studies Quarterly*. 10(2), pp. 111–28.

Bellinzona, C., 2007. Le nozze coi fichi secchi. *Limes*, 3, pp. 107–14.

Berger T. U., 1998. *Cultures of Antimilitarism.* Baltimore: The Johns Hopkins University Press.

Bertolini, M., 2009. Afghanistan, una opportunità strategica per l'Italia. *Rivista Militare*, 3, pp. 46–55.

Biddle, S., 2004. *Military Power: Explaining Victory and Defeat in Modern Battle.* Princeton: Princeton University Press.

Bonvicini G., Carati, A., Colombo, A., Matarazzo, R., Silvestri, S., 2011. La politica estera dell'Italia a 150 anni dall'Unità: continuità, riforme e nuove sfide. In: Alessandro Colombo and Gianni Bonvicini, eds., *La politica estera dell'Italia.* Bologna: Il Mulino.

Bradol, J. H., 2004. *The Sacrificial International Order and Humanitarian Action. In the Shadow of 'Just Wars': Violence, Politics, and Humanitarian Action.* Ithaca, NY: Cornell University Press.

Brighi, E., 2013. *Foreign Policy, Domestic Politics and International Relations.* New York: Routledge.

Brown, M.E., 2014. NATO's Biggest Mistake. *Foreign Affairs*, 8 May, available at: http://www.foreignaffairs.com/articles/141404/michael-e-brown/natos-biggest-mistake.

Brustlein, C., 2011, *Apprendre ou disparaître? Le retour d'expérience dans les armées occidentales.* Focus stratégique n° 33, Paris: IFRI-Laboratoire de Recherche sur la Défense.

Burke, J., 2008. Ten French soldiers killed in Afghanistan as Taliban attacks grow more audacious. *The Guardian*, 20 August.

Buzan, B. and Hansen, L. 2010. Beyond the evolution of international security studies? *Security Dialogue*, 41(6), pp. 659–67.

Calossi, E., Calugi, F. and Coticchia, F., 2013. Peace and war in political discourse of Italian Marxist and post-Marxist parties. *Contemporary Italian Politics*, 5 (3), pp. 309–24.

Cameron, A., 2012. The Channel Axis: France, the UK and NATO. In: Adrian Johnson and Saqeb Mueen (eds). *Short War, Long Shadow The Political and Military Legacies of the 2011 Libya Campaign.* RUSI, pp. 15–24.

Cappelli, R., 2005. Iraq: Italian lessons learned. *Military Review*, 85(2), pp. 58–9.

Catignani, S., 2012. Getting COIN' at the tactical level in Afghanistan: Reassessing counterinsurgency adaptation in the British Army. *Journal of Strategic Studies*, 35(4), pp. 513–39.

Catignani, S., 2014. Coping with knowledge: Organizational learning in the British Army? *Journal of Strategic Studies*, 37(1), pp. 30–64.

Catino M. and Patriotta, G., 2013. Learning from errors: cognition, emotions and safety culture in the Italian Air Force. *Organization Studies*, 34(4), pp. 437–67.

Ceccorulli M. and Labanca, N., eds., 2014. *The EU, Migration and the Politics of Administrative Detention.* New York, NY: Routledge.

Cenciotti, D., 2011. Operation Unified Protector (was Odyssey Dawn) explained: Final Report. *The Aviationist*, available at: http://theaviationist.com/2011/10/26/libya-final-report/.

Chalmers, M., 2010. Unbalancing the force: Prospects for UK defence after the SDSR. *RUSI Future Defence Review Working Paper.*

Chan, S., 1995. Grasping the peace dividend: Some Propositions on the conversion of swords into plowshares. *Mershon International Studies Review*, 39(1), pp. 53–95.

Chauduri, R and Farrell, T., 2011. Campaign disconnect: operational progress and strategic obstacles in Afghanistan, 2009–2011. *International Affairs*, 87, pp. 271–96.

Chiapperini, L., 2012. ISAF 2012. *Rivista Militare*, 4, pp. 101–4

Chiarini, G. M., 2006. I combattimenti urbani nelle CRO'S. *Rivista Miliare*, 3, pp. 46–53.

Ciampini, M., 2010. Intervista al Capo di Stato Maggiore dell'Esercito Gen. C.A. Giuseppe Valotto. *Rivista Militare*, 6, pp. 5–21.

Clarke, M., 2011, Accidental heroes: Britain, France and the Libya Operation – RUSI Interim Libya Campaign Report. *RUSI News*, available at: http://www.rusi.org/news/ref:N4E7B610E8D672/#.UZSri5UbtFh

Cohen, E.A., 1996. A revolution in warfare. *Foreign Affairs*, 75(2), pp. 37–54.

Cohen, W. and Levinthal, D., 1990. Absorptive capacity: A new perspective on learning and innovation, *Administrative Science Quarterly*, 35(1), pp. 128–52.

Cordesman A.H., and Vira V., 2011. The Libyan Uprising: An uncertain trajectory. *Centre for Strategic and International Studies* (CSIS), available at: http://www.google.it/search?q=The+Libyan+Uprising%3A+an+uncertain+trajectory%E2%80%99&i.e.=utf-8&oe=utf-8&aq=t&rls=org.mozilla:it:official&client=firefox-a.

Cordesman, A., 2012. The war in Afghanistan at the end of 2012. *Centre for Strategic and International Studies* (CSIS), available at: http://csis.org/files/publication/130122_AfghanViolence2012_UncertainCourse.pdf

Cornish P., and Dorman, A.M., 2009. National defense in the age of austerity. *International Affairs*, 85(4), pp. 733–53.

Cornish P., and Dorman, A.M., 2012. Rethinking UK national strategy beyond Afghanistan. *International Affairs*, 88(2), pp. 213–22.

Cosentino, M., 2010. Dal Garibaldi al Cavour: L'evoluzione della portaerei italiana. *Rivista Marittima*, 7, pp. 43–56.

Coste, P., 2009. Counterinsurgency and the stabilization phase. In: AA.VV. *The Commitment of Land Forces in Afghanistan 2009 – Doctrine General military Review*, Centre de Doctrine d'Emploi des Forces, pp. 14–16.

Cote, O., 2003. *The Third Battle: Innovation in the U.S. Navy's Silent Cold War Struggle with Soviet Submarines*. Naval War College Newport Papers, 16.

Coticchia, F., 2011. The Enemy at the Gates? Assessing the European military contribution to the Libyan War. *Perspectives on Federalism*, 3(3), pp. 48–69.

Coticchia, F., 2013. *Qualcosa è cambiato. L'evoluzione della politica di difesa italiana dall'Iraq alla Libia (1991–2011)*. Pisa: Pisa University Press.

Coticchia, F., 2014a. *La guerra che non c'era. Opinione pubblica e interventi militari italiani dall'Afghanistan alla Libia*. Milano: UBE Egea.

Coticchia, F., 2014b. A remarkable evolution. Italy's participation in PKOs: Figures and trends. In: A. de Guttry, E. Sommario and L. Zhu, eds, *China's and Italy's Participation in Peacekeeping Operations Existing Models, Emerging Challenges*. Lanham: Lexington, pp. 9–34.

Coticchia, F., and Moro, F.N., 2014. Transforming the Italian Armed Forces, 2001–2013. *The International Spectator*, 49(1), pp. 133–48.

Coticchia, F., Giacomello, G., 2007. In harm's way, why and when a modern democracy risks the lives of its uniformed citizens? *European Security*, 16(2), pp. 163–82.

Coticchia, F., Giacomello, G., 2009. Helping hands, civil-military cooperation and Italy's military operation abroad. *Small Wars and Insurgency*, 20(4), pp. 592–610

Croci, O. and Valigi, M., 2013. Continuity and change in Italian foreign policy: the case of international intervention in Libya. *Contemporary Italian Politics*, 5(1), pp. 38–54.

Davidson, J. W., 2011. *America's Allies and War: Kosovo, Afghanistan and Iraq*. New York: Palgrave Macmillan.

Davidson, J.W., 2013, France, Britain and the intervention in Libya: An integrated analysis. *Cambridge Review of International Affairs*, 26(2), pp. 310–29.

De Lorenzo, C. and Fratini, F., 2008. L'esternalizzazione dei trasporti militari, *Rivista Militare*, 2, pp. 50–69.

Defense News, 2011, *Italy removes aircraft carrier from Libyan campaign*, 7 July 2011.

Del Panta, M. 2014. Mare Nostrum non basta. *Affari Internazionali*, available at: http://www.affarinternazionali.it/articolo.asp?ID=2641

Della Longa, T. and Giuli, A., 2008. Otto mesi di pax italiana in Libano. Novembre 2006-Giugno 2007. *Informazioni Difesa*, 1, pp. 24–7.

Di Feo, G., 2008. Partita Tripla, *L'Espresso*, 10 January, pp. 54–5.

Di Maggio P. and Powell, W.W., 1983. The Iron Cage revisited: Institutional isomorphism and collective rationality in organizational fields. *American Sociological Review*, 48(2), pp. 147–60.

Di Paola, G., 2012. L'Evoluzione della Difesa negli Ultimi Trenta Anni. *RID-Rivista Italiana Difesa*, 10, pp. 26–9.

Doughty, R.A., 1988. The French armed forces, 1918–40. In: Alan Millett and Williamson Murray, eds, *Military Effectiveness, Volume II: The Interwar Period*. London: Allen & Unwin, pp. 39–69.

Downie, R.D., 1998. *Learning from Conflict: The U.S. Military in Vietnam, El Salvador, and the Drug War*, Westport, CT: Praeger.

Downs, A., 1967. *Inside Bureaucracy*, New York, NY: Little & Brown Company.

Dyson T., 2008. Convergence and divergence in post-Cold War British, French, and German military reforms: Between international structure and executive autonomy. *Security Studies*, 17(4), pp. 725–74.

The Economist, 2009. Britain's armed forces losing their way? 29 January.

Egnell, R., 2011. Lessons from Helmand: Afghanistan. What now for British counterinsurgency. *International Affairs* 87(2), pp. 297–315

Engel, G., 1994. Cruise Missiles and the Tomahawk. In: Bradd Hayes and Douglas South, eds, *The Politics of Naval Innovation*, Newport, RI: Naval War College.

Evangelista, M., 1986. Case Studies and Theories of the Arms Race, *Bulletin of Peace Proposals*, 17(2), pp. 197–206.

Evangelista, M., 1988. *Innovation And The Arms Race: How The United States And The Soviet Union Develop New Military Technologies*. Ithaca, NY: Cornell University Press.

Farrell, T., 2010. Improving in War: Military Adaptation and the British in Helmand Province, Afghanistan, 2006–2009. *Journal of Strategic Studies*, 33(4), pp. 567–94.

Farrell, T., 2001. Transnational Norms and Military Development: Constructing Ireland's Professional Army. *European Journal of International Relations*, 7(1), pp. 63–102

Farrell, T., 2008. The dynamics of British military transformation. *International Affairs*, 84(4), pp. 777–807.

Farrell, T. and Gordon S., 2009. COIN machine: the British military in Afghanistan. *RUSI Journal*, 154(3), pp. 18–25.

Farrell, T., Rynning, S. and Terriff, T., 2013. *Transforming Military Power since the End of the Cold War: The United States, Britain, and France, 1991–2012*. Cambridge: Cambridge University Press.

Forte, S. and Marrone, A., 2012. *L'Italia e le missioni internazionali*. Rome: Istituto Affari Internazionali.

Foucault, M., 2012, The Defense Budget in France: Between Denial and Decline. *Focus stratégique n° 36 bis*, Paris: IFRI-Laboratoire de Recherche sur la Défense.

Foust, J. 2012. How France lost Afghanistan. *PBS*, 15 June.

Freedman, L., 2006. The Transformation of Strategic Affairs, *Adelphi Papers 379*, London: International Institute of Strategic Studies.

Gaiani, G., 2007. *Iraq-Afghanistan guerre di pace italiane*. Venezia: Studio LT2.

Giacomello, G., Verbeek B., eds., 2001. *Italy's foreign policy in the 21st Century. The new assertiveness of an aspiring Middle Power?* Lanham, MD: Lexington Books.

Girault, H., 2012. Warfare Doctrine. An ever evolving doctrine. In: AA.VV. T*he Commitment of Land Forces in Afghanistan 2009 – Doctrine General military Review*, Centre de Doctrine d'Emploi des Forces, pp. 19–21.

Global Security, 2013. *Camp Salerno*. 18 March, available at: http://www.globalsecurity.org/military/facility/fob_salerno.htm

Goldman, E.O., 2002. The spread of Western military models to Ottoman Turkey and Meiji Japan. In: T. Farrell and T. Terriff, eds, *The Sources of Military Change: Culture, Politics, Technology*. Boulder, CO: Lynne Rienner, pp. 41–68.

Goldman, E.O., and Andres, R.B., 1999. Systemic effects of military innovation and diffusion. *Security Studies*, 8(4), pp. 79–125.

Goldman, E.O., and Ross, A., 2003. Conclusion. The diffusion of military knowledge: Theory and practice. In E.O. Goldman and L. Eliason, eds, *The Diffusion of Military Technologies and Ideas*. Stanford, CA: Stanford University Press.

Gray, C.S., 2007. Out of the wilderness: Prime time for strategic culture. *Comparative Strategy*, 26(1), pp. 1–20.

Graziano, C., 2013. Un nuovo anno per un nuovo esercito. *Rivista Militare*, 1, pp. 4–13.

Greve, H., 2003. *Organizational Learning from Performance Feedback. A Behavioral Perspective on Innovation and Change*. Cambridge: Cambridge University Press.

Greve, H., 2005. Interorganizational learning and heterogeneous social structure. *Organization Studies*, 26(7), pp. 1025–47.

Griffin, S., 2011. Iraq, Afghanistan and the future of British military doctrine: from counterinsurgency to Stabilization. *International Affairs*, 87(2), pp. 317–33.

Grissom, A., 2006. The future of military innovation studies. *Journal of Strategic Studies*, 29(5), pp. 905–34.

Haddad, S., 2010. Teaching diversity and multicultural competence to French peacekeepers. *International Peacekeeping*, 17(4), pp. 566–77.

Halberstam, D., 2001. *War in a Time of Peace: Bush, Clinton, and the Generals*. New York, NY: Simon & Schuster.

Hallams, E. and Schreer, B., 2012. Towards a post American alliance? *International Affairs*, 88(2), pp. 313–27.

Halperin, M., 1974. *Bureaucratic Politics and Foreign Policy*. Washington, DC: Brookings Institution Press.

Hatch, M.J., 1993. The dynamics of organizational culture. *Academy of Management Review*, 18(4), pp. 657–93.

Hatto, R. 2009. UN command and control capabilities: Lessons from UNIFIL's strategic military cell. *International Peacekeeping*, 16(2), pp. 186–98.

Hotier, G. 2002. *Le 1er RHC dans le Golfe*. Cahiers de RETEX n. 7, Paris: Armée de Terre, CDEF.

House of Commons Library, 2008. *British Defence Policy Since 1997*. Research Paper 08/57, available at: http://ukbriefingpapers.co.uk/briefingpaper/RP94–101.

Huntington, S.P., 1968. *Political Order in Changing Societies*. New Haven, CT: Yale University Press.

IAI-CIRCaP, 2014. *Gli Italiani e la politica estera*. Available at: http://www.iai.it/pdf/Rapporto_IAI-Circap_131217.pdf.

Ignazi P., Giacomello G. and Coticchia F., 2012. *Just Don't Call it War. Italian Military Missions Abroad*. Houndmills: Palgrave Macmillan.

Iron, R., 2013. The Charge of the Knights. *The RUSI Journal*, 158(1), pp. 54–62.

Jacobides, M.G., 2005. The inherent limits of organizational structure and the unfulfilled role of hierarchy: Lessons from a Near-War. *Organization Science*, 18(3), pp. 455–77.

Johnson A. and Mueen, S. (eds) 2012. *Short War, Long Shadow: The Political and Military Legacies of the 2011 Libya Campaign*. RUSI, available at: https://www.rusi.org/downloads/assets/WHR_1–12.pdf.

Johnston, A., 1995 Thinking about strategic culture. *International Security*, 19(4), pp. 32–64.

Kaldor, M. and Salamon, A., 2006. Military force and European strategy. *Survival*, 48(1), pp. 19–34.

Katzenstein, P., 1996. *The Culture of National Security. Norms and Identity in World Politics*. Ithaca, NY: Cornell University Press.

Keaten, J., 2008. France weighing Afghanistan role. *The Associated Press*, 24 August.

Kier, E., 1995. Culture and military doctrine: France between the Wars. *International Security*, 19(4), pp. 65–93.

Kilcullen, D., 2009. *The Accidental Guerrilla: Fighting Small Wars in the Midst of a Big One*. New York and London: Oxford University Press.

King, A., 2010. Understanding the Helmand campaign: British military operations in Afghanistan. *International Affairs*, 86(2), pp. 311–32.

King, A., 2011. Military command in the last decade. *International Affairs*, 87(2), pp. 377–96.

Kometer M.W., and Wright, S.E., 2013. Winning in Libya: By design or default? *IFRI – Focus stratégique*, 41, available at: http://www.ifri.org/?page=contribution-detail&id=7516.

Lampel, J., Shamsie, J. and Shapira, Z., 2009. Experiencing the improbable: Rare events and organizational learning. *Organization Science*, 20(5), pp. 835–45.

Lewis, J.A.C., 2002. French fighters join action in Afghanistan. *Jane's Defence Weekly*, March 13.

Little, P., 2009. Lessons unlearned. *RUSI Journal*, 154(3), pp. 10–16.

Locatelli, A., 2014. *The Italian Defence Policy after the Cold War. Europeanisation, American Influence, or Strategic Adjustment?* unpublished manuscript.

Luttwak, E., 1996. A post-heroic military policy. *Foreign Affairs*, 75(4), pp. 33–44.

Mansoor, 2009. The British army and the lessons of the Iraqi war. *British Army Review*, 147, pp. 11–15.

March, J., 1991. Exploration and exploitation in organizational learning. *Organization Science*, 2(1), pp. 71–87.

March, J., 2010. *Rediscovering Institutions*. New York, NY: Simon & Schuster.

Marchi Balossi-Restelli, L. ed., 2011. *A EU Innovative External Action?* Cambridge: Cambridge Scholars Publishing.

Marrone, A., 2012. La crisi abbatte i bilanci della difesa, *Affari Internazionali*, 26 July, available at: http//www.affarinternazionali.it/articolo.asp?ID=2095.

Marston, D., 2009. 'Smug and complacent'? Operation Telic. *British Army Review*, 147, pp. 16–23.

Mason, M., 2007. Libano: verso quale futuro? *RID-Rivista Italiana Difesa*, 6, pp. 34–42.

Mason, M., 2013. Difesa 2013: Eppur si muove. *RID-Rivista Italiana Difesa*, 1, pp. 22–35.

Mearsheimer, J., 1983. *Conventional Deterrence*. Ithaca: Cornell University Press

Mearsheimer, J., 1990. Back to the future: Instability in Europe after the Cold War. *International Security*, 15(1), pp. 5–56.

Mérand, F., 2008. *European Defence Policy. Beyond the Nation State*. Oxford: Oxford University Press.

Miranda V., 2011. Striking a balance between norms and interests in Italian Foreign Policy: The Balkans and Libya. *IAI Working Papers*, 11.
Mounier, G., 2009, Civilian crisis management and the external dimension of JHA: Inceptive, functional, and institutional similarities. *Journal of European Integration*, 31, pp. 45–64.
Murray, W. and Millett, A., eds, 1996. *Military Innovation in the Interwar Period.* Cambridge: Cambridge University Press.
Murray, W., 1996. Innovation: Past and future. In: Alan Millett and Williamson Murray, eds, *Military Innovation in the Interwar Period*, Cambridge: Cambridge University Press.
Nagl, J., 2005. *Learning to Eat Soup with a Knife. Counterinsurgency Lessons from Malaya and Vietnam*. Chicago, IL: Chicago University Press.
Nones, M. and Marrone, A., 2011. *La Trasformazione delle Forze Armate: il programma Forza NEC*. Roma: Quaderni IAI.
Olsen, J.A., 2007. *John Warden and the Renaissance of American Air Power*, Baltimore, MD: Potomac Books.
Osservatorio di Politica Internazionale, 2010. *Rapporto Collettivo: Le missioni internazionali*. Available at: http://www.iai.it/pdf/Oss_Polinternazionale/pi_r_0004.pdf.
Panebianco, A., 1997. *Guerrieri democratici*. Bologna: Il Mulino.
Panorama, 2011. *La guerra italiana in Libia: ecco le prime cifre ufficiali*. 3 November.
Patrick, S., 2011. Libya and the future of humanitarian intervention. How Qaddafi's fall vindicated Obama and RtoP. *Foreign Affairs*, available at: http://www.foreignaffairs.com/articles/68233/stewart-patrick/libya-and-the-future-of-humanitarian-intervention?page=show.
Petrilli, L. and Sinapi, V., 2007. *Nassiriya, la vera storia*. Torino: Lindau.
Pierson, P., 1994. *Dismantling the Welfare State. Reagan, Thatcher, and the Politics of Retrenchment*. Cambridge: Cambridge University Press.
Pierson, P., 2004. *Politics in Time*. Princeton, CA: Princeton University Press.
Posen, B.R., 1984. *The Sources of Military Doctrine. France, Britain and Germany between the World Wars*. Ithaca, NY: Cornell University Press.
Posen, B.R., 2006. European Union security and defense policy: Response to unipolarity? *Security Studies*, 15(2), pp. 149–86.
Posen, B.R. and Arnold, J. 2004. ESDP and the structure of World Power. *The International Spectator*, 39(1), pp. 5–17.
Pressman, J. and Wildavsky, A., 1984, *Implementation How Great Expectations in Washington Are Dashed in Oakland; Or, Why It's Amazing that Federal Programme Work at All, This Being a Saga of the Economic Development Administration as Told by Two Sympathetic Observers Who Seek to Build Morals on a Foundation*. Berkeley: UC Press.
Quintana, E. 2012. The war from the air. In: Adrian Johnson and Saqeb Mueen, eds, *Short War, Long Shadow. The Political and Military Legacies of the 2011 Libya Campaign*. RUSI, pp. 31–40.

Quintana, E., 2011, The air operation. In: AA.VV, *Accidental Heroes. Britain, France and the Libya Operation, An interim RUSI Campaign Report*, pp. 5–7, available at: http://www.rusi.org/news/ref:N4E7B610E8D672/.

Ratti, L., 2011. Italy as a multilateral actor: The inescapable destiny of a middle power. In: Maurizio Carbone, ed., *Italy in the Post-Cold War Order.* Lanham: Lexington Books, pp. 123–40.

Resende-Santos, J., 2007. *Neorealism, States, and the Modern Mass Army*. New York and Cambridge: Cambridge University Press.

Ricks, T. E., 2009. *The Gamble. General David Petraeus and the American military adventure in Iraq, 2006–2008*. New York, NY: Penguin Group Press.

Ringsmose, J. and Børgesen B., 2011. Shaping public attitudes towards the deployment of military power: NATO, Afghanistan and the use of strategic narratives. *European Security*, 20(4), pp. 505–28.

Rivista Aeronautica, 2012. *Missione Libia 2011 – Il Contributo dell'Aeronautica Militare*. Roma: Aeronautica Militare – Edizione Rivista Aeronautica.

Rosen, S.P., 1988. New ways of war: Understanding military innovation. *International Security*, 13(1), pp. 134–68.

Rosen, S.P., 1991. *Winning the Next War: Innovation and the Modern Military*. Ithaca, NY: Cornell University Press.

Rosenberg, B. 2010. Battlefield network connects allied forces in Afghanistan. *Defence Systems*, 14 September, available at: http://defencesystems.com/articles/2010/09/02/c4isr-2-afghan-mission-network-connects-allies.aspx.

Russell, J., 2011. *Innovation, Transformation, and War. Counterinsurgency Operations in Anbar and Ninewa Provinces, Iraq, 2005–2007*. Stanford, CA: Stanford University Press.

Rynning, S. 2003. The European Union: Towards a strategic culture? *Security Dialogue*, 34(4), pp. 479–96.

Saideman, S. M. and Auerswald, D. P., 2012. Comparing caveats: Understanding the sources of national restrictions upon NATO's mission in Afghanistan. *International Studies Quarterly*, 56(1), pp. 67–84.

Santoro, C.M. 1991. *La politica estera di una media potenza. L'Italia dall'Unità ad oggi*. Bologna: II Mulino.

Sapolsky, H., 1972. *The Polaris System Development*. Cambridge. MA: Harvard University Press.

Schein, E.H., 2004. *Organizational Culture and Leadership*. San Francisco, CA: Wiley and Co.

Schmitt, G. J., 2012. Italian hard power: Ambitions and fiscal realities. *American Enterprise Institute – National Security Outlook*, 3, pp. 1–11.

Serra, P., 2012. Il commando dell'operazione UNIFIL. *Rivista Militare*, 4, pp. 88–91.

Shapiro, J., 2002. The role of France in the war on terrorism. *Brookings Institute*, available at: http://www.brookings.edu/fp/cusf/analysis/shapiro.pdf.

Simoni, S., 2013. *Understanding Transatlantic Relations: Whither the West?* New York, NY: Routledge.

Smith, R., 2006. *The Utility of Force*. London: Penguin.

Stone, D., 2004. Transfer agents and global networks in the 'transnationalization' of policy. *Journal of European Public Policy*, 11(3), pp. 545–66.

Strachan, H. 2011. The strategic gap in British defence policy. *Survival*, 51(4), pp. 49–70.

Thomas, J., Watts Sussman, S. and Henderson, J., 2001. Understanding 'Strategic Learning': Linking organizational learning, knowledge management, and sensemaking, *Organization Science*, 12(3), pp. 331–45.

Townend, M., 2009. Troops say they lack the right kit to fight in Helmand. *The Guardian*, 9 August.

Ucko, D., 2008. Innovation or inertia: The U.S. military and the learning of counterinsurgency, *Orbis*, 52(2), pp. 290–310.

Ucko, D., 2010. Lessons from Basra: The future of British counterinsurgency, *Survival*, 52(4), pp. 131–58.

Vennesson P. et al., 2009. Is there a European way of war? Role conceptions, organizational frames, and the utility of force. *Armed Forces and Society*, 35(4), pp. 628–45.

Vetere, F. 2010. Le cluster bombs e la bonifica umanitaria in Libia. *Rivista Militare*, 4, pp. 3–5.

The Wall Street Journal, 2011. Transcript of Defence Secretary Gates's speech on NATO's future. June 12, available at http://on.wsj.com/k0ODZr.

Waltz, K.N., 1979. *Theory of International Politics*, Reading: Addison-Wesley.

Wendling, C., 2010. *The Comprehensive Approach to Civil-Military Crisis Management. A Critical Analysis and Perspective*. Paris: IRSEM.

Whitman, R. G., 2004. NATO, the EU and ESDP: An emerging division of labour? *Contemporary Security Policy*, 25(3), pp. 430–51.

Willet, L., 2011. The maritime contribution. In: AA.VV, *Accidental heroes: Britain, France and the Libya Operation – RUSI Interim Libya Campaign Report*, available at: http//www.rusi.org/news/ref,N4E7B610E8D672/.

Official Documents

Commissione Difesa, 2013. *Indagine conoscitiva sui sistemi d'arma destinati alla difesa in vista del Consiglio Europeo di Dicembre 2013*. Defence Commission, Chamber of Deputies, 19 March 2014.

Department of Defense, DoD, 2001. *Quadriennal Defense Review*. Available at: http://www.defense.gov/pubs/qdr2001.pdf.

European Council – EUCO 2013, 79/13. *Defence in Europe: Pragmatically Forward*. Speech by President of the European Council Herman Van Rompuy at the annual conference of the European Defence Agency 'European Defence Matters'.

Ministère de Finances, 2010. *Mission ministérielle annexe à la loi de finances initiale pour 2010*. Available at: http://www.performance-publique.budget.

gouv.fr/sites/performance_publique/files/farandole/ressources/2010/lfi/pdf/DBGNORMALLFIMSNDA.pdf.
Ministère de la Défense, 1972. *Livre Blanc de la Défense*. Paris.
Ministère de la Défense, 1984. *Livre Blanc de la Défense*. Paris.
Ministère de la Défense, 2008. *Livre Blanc de la Défense*. Paris.
Ministère de la Défense, 2010. *Concept d'Emploi des Forces – CEF*. Paris.
Ministère de la Défense, 2014. Annuaire Statistique de la Défense, years 2001–2013, Paris.
Ministero della Difesa, 1977. *Libro Bianco della Difesa*. Roma: Stato Maggiore della Difesa.
Ministero della Difesa, 1985. *Libro Bianco della Difesa*. Roma: Stato Maggiore della Difesa.
Ministero della Difesa, 1991. *Modello di difesa. Lineamenti di sviluppo delle FF.AA. negli anni '90*. Roma: Stato Maggiore della Difesa.
Ministero della Difesa, 2001. *Nuove Forze per un Nuovo Secolo*. Roma: Stato Maggiore della Difesa.
Ministero della Difesa, 2005. *Concetto Strategico del Capo di Stato Maggiore*. Roma: Stato Maggiore della Difesa.
Ministero della Difesa, 2005. *Investire in Sicurezza*. Roma: Stato Maggiore della Difesa.
Ministero della Difesa, 2005. *La Trasformazione Net-Centrica. Il Futuro dell'Interoperabilità Multinazionale e Interdisciplinare*. Roma: Stato Maggiore della Difesa.
Ministero della Difesa, 2012. *Approccio Nazionale Multi-Dimensionale alla gestione delle crisi*. Documento di riflessione congiunto Ministeri Esteri e Difesa. Roma.
Ministero della Difesa, 2013. *Direttiva Ministeriale in merito alla politica militare per l'anno 2013*, Roma.
Ministry of Defense, 1990. *Options for Change*. London: Whitehall.
Ministry of Defense. 1994. *Frontline First: The Defense Costs Study*. London: Whitehall.
Ministry of Defense, 1998. *The Strategic Defense Review*. London: Whitehall.
Ministry of Defense, 2002. *New Chapter of Strategic Defense Review*. London: Whitehall.
Ministry of Defense, 2003. *White Paper*. London: Whitehall.
Ministry of Defense, 2004. *Delivering Security in a Changing World*. London: Whitehall.
Ministry of Defense, 2010. *Strategic Defense and Security Review*. London: Whitehall.
Ministry of Defence, 2012. *The UK's contribution to freeing Libya*. available at: http://www.mod.uk/DefenceInternet/DefenceNews/MilitaryOperations/TheUksContributionToFreeingLibya.htm.
NATO-NSA, 2008. AAP-006. *NATO Glossary of terms and definitions*.
NATO, 2009. *JDP 3–40 – Security and Stabilisation: the military contribution*.

NATO, 2010. *Active Engagement, Modern Defence. Strategic Concept for the Defence and Security of the Members of the North Atlantic Treaty Organization.* Lisbon, 19–20 November.

Partito Democratico (Democratic Party – PD), 2014. *Considerazioni conclusive sull'indagine conoscitiva sui sistemi d'arma,* Defence Commission, Chamber of Deputies, 19 March.

Servizio Studi Camera dei Deputati, 2010. *Nuovi profili della partecipazione italiana alle missioni militari internazionali.* Available at: http://documenti.camera.it/leg16/dossier/Testi/di0238.htm#_Toc268187872.

Solana, J., 2003. *European Security Strategy – ESS (2003), A Secure Europe in a Better World.* Brussels, 12 December.

Stato Maggiore dell'Aeronautica, SMA, 2012, Aeronautica Militare. Direzione per l'Impiego del Personale Militare dell'Aeronautica. *Direttiva per l'impiego degli ufficiali, sottufficiali graduati e militari di truppa in ambito nazionale ed internazionale.* Edizione 2012, DIPMA, U.D. 001.

Stato Maggiore dell'Esercito, SME, 2010. *Army Report 2010.*

Stato Maggiore dell'Esercito, SME, 2011. *Army Report 2011.*

Stato Maggiore dell'Esercito, SME, 2012. *Army Report 2012.*

Stato Maggiore dell'Esercito, SME, 2011. *Linee Guida e Direttive sulla formazione e l'impiego del Personale Militare. Edizione 2011.*

Stato Maggiore della Difesa, SMD, 2009. *Glossario dei termini e* delle definizioni SMD-G-024.

Stato Maggiore della Difesa, SMD, 2011. III Reparto Centro Innovazione della Difesa. *La Dottrina Militare Italiana.* PID/S-1.

Stato Maggiore della Difesa, SMD, 2012. *Military Implications of the Future Operational Environment.* 3rd Division – Military Policy and Planning Centre for Defence Innovation. Capstone Concept 001 (CC001).

Stato Maggiore della Difesa, SMD, III REPARTO Centro Innovazione della Difesa, 2012. *La Protezione delle Forze.* Pubblicazione Interforze Dottrinale (PID)/O-3.14.

Stato Maggiore della Difesa, SMD, III REPARTO Centro Innovazione della Difesa, 2013. *Direttiva di policy in materia di Lezioni Identificate/Lezioni Apprese.* SMD-G-027(A).

Stato Maggiore della Difesa, SMD, 2013. *Direttiva di policy in materia di Lezioni Identificate/ Lezioni Apprese* – SMD-G-027(A).

US Army and Marine Corps Field Manual, 2007. *FM 3–24,* available at:http://fas.org/irp/doddir/army/fm3–24fd.pdf.

Appendix

List of Interviews

Officer A, Folgore Parachute Brigade, field experience in Afghanistan (18 February 2013)
Officer B, Folgore Parachute Brigade, field experience in Afghanistan (28 February 2013)
General Commander, Folgore Parachute Brigade (17 July 2013)
Étienne de Durand, Institut Français des Relations Internationales – IFRI (6 November 2013)
Corentin Brustlein, Institut Français des Relations Internationales – IFRI (6 November 2013)
Guillaume Garnier, Lt.Col., Armée de Terre and IFRI (6 November 2013)
Officer, General Staff of the Army, Liaison Office with Armée de Terre (8 November 2013)
Malcolm Chalmers, King's College, London (10 December 2013)
Theo Farrell, King's College, London (10 December 2013)
Trevor Taylor, King's College, London (11 December 2013)
Staff officer A, General Staff of the Army, (22 January 2014)
Staff officer B, General Staff of the Army, (22 January 2014)
Staff officer, NATO, experience as Commander of the Italian Task Force in RC-W (1 February 2014)
Officer, responsible for Lessons Learned, Taurinense Alpine Brigade (5 July 2014)
Officer A, Italian Air Force, Institute of Military Aeronautical Sciences (4 February 2014)
Officer B, Italian Air Force, Institute of Military Aeronautical Sciences (4 February 2014)
Officer C, Italian Air Force, Institute of Military Aeronautical Sciences (4 February 2014)
Officer, NATO Rapid Deployable Corps Italy (5 March 2014)
Staff Officer, General Staff of the Army, (28 April 2014)
General Commander, Taurinense Alpine Brigade (5 July 2014)
Officer, former Commander of a 'Weapon Intelligence Team', Taurinense Brigade (5 July 2014)
General Commander, Pinerolo Mechanized Brigade (24 July 2014)

Major Programmes in Italy

What follows is a list of major programmes divided by sector, with associated costs (sustained by year) according to data coming from the Ministry of Defence (mostly, the *Nota Aggiuntiva* to the defence budget), the Ministry of Economic Development (MiSE).

Table A.1 Joint and Army Major Programmes, 2001–2011

Year	Joint Programmes		Army Programmes	
	Expenditures (,000 €)	Programme Name	Expenditures (,000 €)	Programme Name
2001	n/a	n/a	479	ARIETE 1 [*] DARDO [**]
2002	n/a	n/a	129	DARDO [**]
2003	107	F-35 JSF [*]	245	NH-90 [†] DARDO
2004	126	F-35 JSF [*]	189	NH-90 [†]
2005	160	COSMO-SKYMED [**] F-35 JSF [*]	215	NH-90 [†]
2006	167	TLC *Interforze* [†] F-35 JSF [*]	113	Tactical and Logistical Vehicles, PZH2000 [††]
2007	225	C4I F-35 JSF [*]	667	NH-90 [†] VTLM [§] FSAF [§§] PZH2000 [††]
2008	301	F-35 JSF [*] C4I WIMAX [††]	566	NH-90 [†] FSAF SAMP/T PZH2000 [††]
2009	280	WIMAX C4I	442	PZH2000 [††] VTLM [§] NH-90 [†]
2010	329	F-35 JSF [*] WIMAX	281	NH-90 [†] VTLM [§]
2011	469	F-35 JSF [*]	344	NH-90 [†] ETM [‡]
Legend	[*] Multi-role Fighter [**] Satellites [†] TLC systems for Joint Forces [††] ICT systems for Joint Forces		[*] Main Battle Tank [**] Infantry Fighting Vehicle [†] Multi-role Helicopter [††] Artillery [§] Tactical Light Multi-role Vehicles (*Lince*) [§§] Surface-to-Air Missile Platform/Terrain [‡] Medium-lift Transport Helicopter	

Table A.2 Navy and Air Force Major Programmes, 2001–2011

Year	Navy Programmes		Air Force Programmes	
	Expenditures (,000 €)	**Programme Name**	**Expenditures (,000 €)**	**Programme Name**
2001	699	ORIZZONTE-class frigates Pattugliatori di Squadra [*] Carrier Planes and Helicopters	1547	EUROFIGHTER [*] TORNADO [**] C130J [†]
2002	733	ORIZZONTE-class frigates U-212 [**] Pattugliatori di Squadra [*]	890	EUROFIGHTER [*] TORNADO [**]
2003	465	ANDREADORIA carrier [†] ORIZZONTE-class frigates U-212 [**] PAAMS [††]	875	EUROFIGHTER [*] TORNADO [**] B767
2004	541	ANDREADORIA carrier [†] ORIZZONTE-class frigates U-212 [**] PAAMS [††]	779	C130J [†] EUROFIGHTER [*] TORNADO [**]
2005	548	NH-90 CAVOUR carrier ORIZZONTE-class frigates U-212 [**]	696	B767 [††] EUROFIGHTER [*] TORNADO [**]
2006	106	U-212 [**]	257	B767 [††] EUROFIGHTER [*] TORNADO [**]
2007	587	NH-90 [§] CAVOUR carrier ORIZZONTE-class frigates	737	ALTBM-MEADS TORNADO [**] EUROFIGHTER [*]
2008	443	ORIZZONTE-class frigates CAVOUR carrier NH-90 [§]	410	TORNADO [**] EUROFIGHTER [*]
2009	335	ORIZZONTE-class frigates CAVOUR carrier carrier NH-90 [§]	255	ALTBM-MEADS [§] TORNADO [**] C4I
2010	360	U-212 [**] NH-90 ORIZZONTE-class frigates	184	TORNADO [**]
2011	379	U-212 [**] NH-90 [§]	173	TORNADO [**]
Legend	[*] Frigate-size ships with no ASW capabilities [**] Diesel-Electric Submarine [†] First project name of the Cavour carrier [††] Principal Anti Air Missile System [§] Multi-role Helicopter		[*] Multi-role Fighter [**] Fighter aircraft [†] Transport aircraft [††] Tanker aircraft [§] Missile system	

Index

The Index covers all material in the six chapters of the text. Titles of works are indicated in *italics*. Page numbers appearing in *italics* indicate the presence of a table, figure or other illustration on that page. Cited works are indexed by the name of the lead author (or editor) and then publication date, as given in the text. Full titles are given in the References.